W9-BIP-845

THE *Field Guide* TO RABBITS

THE *Field Guide* TO RABBITS

SAMANTHA JOHNSON

First published in 2008 by MBI Publishing Company and Voyageur Press, an imprint of MBI Publishing Company, 400 1st Avenue North, Suite 300, Minneapolis, MN 55401 USA

The information in this book is true and complete to the best of our knowledge. All recommendations are made without any guarantee on the part of the author or Publisher, who also disclaim any liability incurred in connection with the use of this data or specific details.

We recognize, further, that some words, model names, and designations mentioned herein are the property of the trademark holder. We use them for identification purposes only. This is not an official publication.

Voyageur Press titles are also available at discounts in bulk quantity for industrial or sales-promotional use. For details write to Special Sales Manager at MBI Publishing Company, 400 1st Avenue North, Suite 300, Minneapolis, MN 55401 USA.

To find out more about our books, join us online at www.voyageurpress.com.

Library of Congress Cataloging-in-Publication Data

Johnson, Samantha.
The field guide to rabbits / by Samantha Johnson.
p. cm.
ISBN-13: 978-0-7603-3193-4 (printed laminated cover)
ISBN-10: 0-7603-3193-6 (printed laminated cover) 1. Rabbits—Identification. I. Title.
QL737.L32J64 2008
636.932'2—dc22

2007009184

Editor: Amy Glaser
Designer: Julie Vermeer

Printed in China

All Photgraphy by Daniel Johnson

About the Author: Samantha Johnson is the author of *How to Raise Horses*. She received her first rabbit at age 11 and has been a rabbit owner and enthusiast ever since. Johnson lives in Phelps, Wisconsin.

About the Photographer: Daniel Johnson enjoys photographing animals such as rabbits, dogs, farm animals, gardens, and rural life. His images appear in magazines, books, greeting cards, and calendars nationwide.

Page 8: This Polish rabbit is a Blue-Eyed White.

Facing Page: This mixed-breed bunny goes undercover in the tall grass.

Dedication

To my grandfather, Loyal Behling, for instilling his great love of rabbits in me and for giving me excellent advice on how to have fun with them: "Let them little guys go!"

Acknowledgments

It is a strange feeling to write an acknowledgment page because I am attempting to fill it with thanks to everyone who has helped with a specific project. It's hard to calculate the amount of support, encouragement, and advice that has been given to me over the course of this project, and with great appreciation I specifically thank the following individuals:

My entire family, for putting up with months of rabbit research and papers all over the place. Dan and Norvia, for providing so many super photographs. The Fox Valley Rabbit Club and Dennis Roloff, for their cooperation and enthusiasm. Amy Glaser, my editor, for her assistance and encouragement. My parents, for buying me my first rabbit at age 11 and for helping me upgrade to purebred Netherland Dwarf rabbits later on. Thanks to Rosie, for helping identify colors; Em for photo setup; Josh for locating research materials; and to Linny and Chocie for being so quirky and making me laugh. And a special thanks to Daknees, my favorite rabbit ever. Special acknowledgment to the following photographers who graciously allowed their images to be reproduced: Heather Dunaway (English Spot), Lynne Schultz (Silver), Ursula Glauser (Satin Angora), Shannon Hubner (Beveren), Ed White (American Sable), Franco and Tracy Rios (American and American Chinchilla), Janet Gruber (Giant Angora), Barbi Brown (Lilac), Andrea McAvoy (Blanc de Hotot), Mara Scott (Cinnamon), and Helga Vierich-Drever (Canadian Plush Lop). Many thanks to Candis Hankins and Pamela Nock for answering my questions and reading through portions of the manuscript. To all of these individuals, as well as the many others who have offered assistance and support, many, many thanks!

Facing page and above: Thanks to Josh, Emily, and Anna for their assistance in acquiring models. It was a lot of fun!

Contents

Introduction

Everyone has their own memories about how they became acquainted with rabbits. For many people who grew up in rural areas, rabbits were the ideal 4-H project. They are easy to keep, fun to have, and enjoyable to show. For people who grew up in an urban setting, a rabbit made an ideal small pet. They don't take up much room, don't eat much, and definitely don't bother the neighbors by barking! My first introduction to rabbits was the one I received at age 11, a black doe of unknown background that was a medium-sized bunny with an "average" temperament. She was named Cuddles but was not noted for being particularly cuddly. She promptly produced a litter of kits shortly after I brought her home, and she adopted a cannibalistic attitude toward the innocent youngsters. All four kits survived, albeit with smaller ears than are typically found on newborn kits! I kept one of the kits, Boo Boo (aptly named for the ear mishap; she had none), for her entire life and she never seemed any the worse for not having ears. Later on I added a pair of Netherland Dwarf rabbits; a black Himalayan doe and a Siamese smoke pearl buck. Although I never bred the pair, they were my pets for many years and I very much enjoyed them. Thinking back, I remember them being particularly adorable, perhaps because all I had to compare them to was my no-eared Boo Boo.

Perhaps you've had a similar experience with rabbits, enjoyed them as a child, and now you are looking to re-embrace the cuddly creatures, maybe for your own enjoyment or to share with your children. Or perhaps you're a newcomer looking to purchase a rabbit or begin a small rabbitry. This book will help guide you through the maze of rabbits and arm you with information on all of the breeds recognized by the American Rabbit Breeders Association, as well as provide you with information on their shapes, sizes, colors, fur varieties, and more. In addition, we

Facing page: This Holland Lop rabbit is standing up to see something while exploring outside.

Three rabbits exhibiting different colors display unique characteristics such as broken and solid color patterns, differences in eye color, and ear length.

will acquaint you with the basic history of rabbits; offer insight on bunny behavior; discuss pregnancy, kindling, and raising kits; and explore the fascinating world of rabbit shows.

Whether you are just starting out or already have years of experience, this guide places a wonderful quantity of information at your fingertips so you can quickly identify a rabbit's breed wherever you happen to be: wandering about a county fair, attending a rabbit show, or shopping at a pet shop (or pet swap, depending on your location).

As you'll come to discover, rabbits maintain a huge following of devoted enthusiasts, partly due to their many unique varieties and types, their charming personalities and user-friendly maintenance, and because no one can resist the feeling of a soft bundle of fur in their arms. Hoppy trails!

Facing page: Standing up allows for a better view. This Lionhead-mix rabbit is obviously curious about something.

Down the Rabbit Trail

Rabbits through the Years

Once upon a time, long before there was such a thing as a Silver Marten Netherland Dwarf or a Black Checkered Giant, there were wild rabbits. It is said that there are more than 20 varieties of wild rabbits throughout the world, but only one variety of European rabbit (*Oryctolagus cuniculus*) is the foundation of today's domestic rabbits. Rabbits were traded by the Phoenicians as they traveled, as the small animals were noted as a source of meat and fur. The first domestication of rabbits dates back to the Middle Ages when French Catholic monks began breeding rabbits and eventually began focusing on different coat color patterns. By the 1500s, rabbits were becoming more popular as domestic farm animals and more varieties in color, size, and eventually breeds, were being established. Selective breeding for a gentler and quieter temperament was also being conducted. In addition to the domestication that took place in Europe, rabbits have been bred by the Chinese for centuries.

One of the earliest known breeds of domestic rabbit is the English Lop. Although the origin of this breed has never been conclusively reached, it possibly originated in North Africa sometime before the 1850s, and possibly prior to the year 1700. History shows that English Lops were being exhibited in England at shows in the late 1800s. Other early domesticated rabbit breeds include the Champagne d'Argent (developed in France) and the Dutch (developed in Holland). The first domesticated rabbits were brought to the United States in the early 1700s, and some of the earliest imported breeds were the Angora, Dutch, and Polish. There was also considerable interest in the Belgian Hare in the early years of rabbit breeding in the United States, with the historic Belgian Hare Boom occurring between 1898 and 1901.

Facing page: Rabbits have been popular as children's pets for more than a century. This French Angora makes an excellent tea party friend for a little girl.

In the Victorian era of elegance, grace, and sophistication, animal fancy shows began to gain in popularity, and the rabbit's image began a transformation into that of a charming household pet rather than a small farm animal kept for meat and fur purposes.

The first all-breed national rabbit organization in America was founded in 1910 and was essentially the predecessor to today's American Rabbit Breeders Association (ARBA). At that time, however, it was known as the National Pet Stock Association. The organization underwent several name changes before the current name was created.

Although not native to either country, wild rabbits have multiplied in their proverbial fashion in both New Zealand and Australia. Rabbits were originally brought to Australia in the 1850s and to New Zealand in 1860. By the late 1800s, the wild rabbit population in both countries had reached epic proportions, with an estimated 20 million rabbits in Australia. This has been a long-term problem in both countries. Because of the extensive agricultural damage caused by the rabbits, efforts have been made to reduce the wild rabbit population. These measures have included the release of predators, such as ferrets and weasel; the use of

Today's domestic rabbits descend from one variety of European rabbit (*Oryctolagus cuniculus*), but there are many other types of wild rabbit throughout the world. This is an Eastern Cottontail (*Sylvilagus floridanus*) spotted in Florida.

Wild rabbits have caused extensive agricultural damage in Australia and New Zealand, prompting both countries to undertake measures to reduce the wild rabbit population.

The Chinese zodiac cycle's Year of the Rabbit is recognized and commemorated by this United States postage stamp from the 1990s. The Year of the Rabbit occurs every 12 years.

guns and poisons; and the release of viruses, such as myxoma and calicivirus. While these methods have achieved some measure of success, they have not eradicated the wild rabbit in either country, although it is said that the rabbit population in Australia has decreased by 90 percent since these measures have been undertaken.

The age-old Chinese zodiac cycle features the Year of the Rabbit once every 12 years, and people born in those years are said to be quiet, sentimental, sensitive, diplomatic, and detail-oriented. They are said to make excellent politicians, though other professions are certainly represented. Some of the more famous individuals born in these years include Frank Sinatra, Bob Hope, Pope Benedict XVI, Albert Einstein, and Tiger Woods. The next Year of the Rabbit will span from February 3, 2011, to January 22, 2012.

The Rabbit Today

The popularity of domestic rabbits continues to grow, and the American Rabbit Breeders Association (ARBA) currently is supported by more than 30,000 members. ARBA-sanctioned shows are enthusiastically attended by members, with more than 3,000 sanctioned shows per year and an average of 850,000 rabbits shown nationally each year. The ARBA annual convention is also the largest show of the year, with more than 20,000 rabbits often in attendance at this single show. The location of the convention rotates each year.

Rabbits We Love

If you think back to your childhood, you're bound to have warm and fuzzy feelings when you recall some of the most familiar rabbits in our culture. Perhaps influenced by the immense popularity of the Beatrix Potter children's books, rabbits have long been a prominent part of children's literature. Who can forget those hours of reading *Peter Rabbit*, *The Velveteen Rabbit*, or *The Runaway Bunny*? Or the delightful tales of *The Country Bunny and the Little Gold Shoes*, *Alice in Wonderland* and the entertaining White Rabbit, or *Goodnight Moon* and its little baby rabbit in the Great Green Room?

Then again, perhaps you fondly remember those carefree Saturday mornings watching the "Wascally Wabbit" himself, Bugs Bunny, as he tricked Elmer Fudd for the umpteenth time. Or maybe you preferred the antics of Rabbit from *Winnie-the-Pooh* and his tendency toward blunt and sometimes misguided advice in A. A. Milne's classic Pooh stories and the subsequent television and film versions. Of course, we cannot forget the allure of the Easter Bunny and the thrill of the Easter egg hunt, or the never-give-up attitude of the Energizer Bunny! In any case, your earliest recollections most likely include one or more of these famous rabbits. When viewed in the context of childhood reminiscences, rabbits seem to represent all that is comforting and adorable and peaceful, such as the sweetness of your mother reading to you; the softness of a plush toy; laughter, giggling, and joy. In short, they are the epitome of childhood.

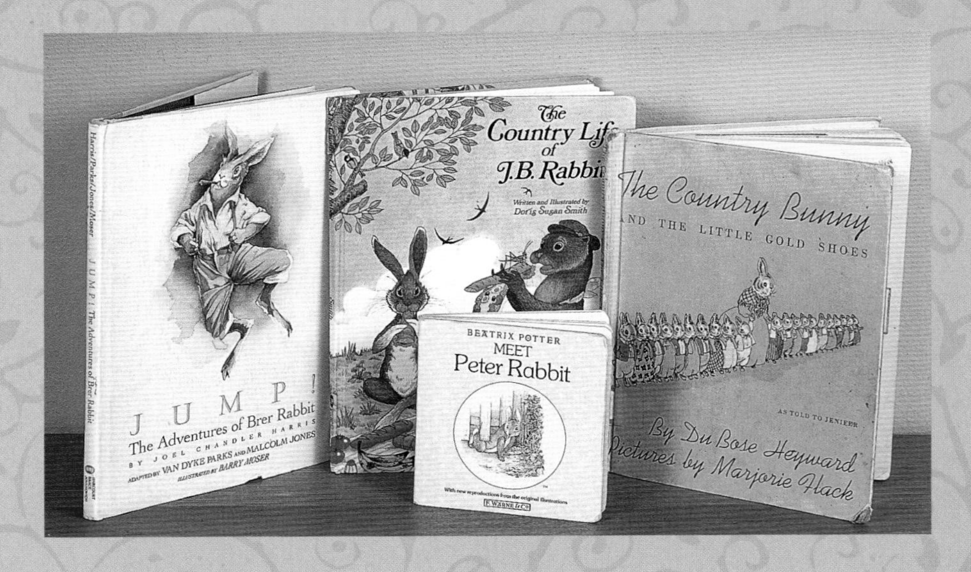

Rabbits have long been featured on collectible items, are popular as children's stuffed animals, and common on decorative home items. The Royal Doulton line of Bunnykins items have excellent collectible value and come in a complete line of plates, cups, saucers, egg cups, piggy banks, and figurines. Rabbits have also been featured on postage stamps worldwide and have appeared on T-shirts, sweatshirts, afghans, needlework kits, stoneware, and dozens of other collectible items. Rabbit enthusiasts can literally decorate their homes to reflect their passion. Then of course, we must not forget one of the symbols of good luck, the rabbit's foot!

The small size of many rabbit breeds makes them ideal pets for a child. They are easy to handle and care for and are an enjoyable friend.

There are nearly 200 breeds of rabbits worldwide, and ARBA recognizes 47 of these. Additional breeds are currently in the process of gaining ARBA approval. Many of the recognized breeds are very popular and well supported by enthusiasts; however, some of the breeds are considered to be quite rare in the United States. The American Livestock Breeds Conservancy (ALBC) is a nonprofit organization founded in 1977 to help preserve and promote rare and endangered livestock breeds. The ALBC currently highlights 11 ARBA-recognized rabbit breeds that have less than 200 annual registrations, including 3 breeds with less than 50 annual registrations. These breeds are on the ALBC's critical, threatened, and watch lists.

Many 4-H members enjoy raising rabbits and showing them as 4-H projects. County fairs typically have extensive exhibits of rabbits shown by youth members as well as by adult enthusiasts.

Today, rabbits are viewed as an all-purpose animal, enjoyed as pets, for breeding, as show animals, and also for their fur and meat. They are enjoyed by children as an ideal first pet. Children experience the responsibility of pet care with an animal that is small and easily handled. They are popular with 4-H members and are often exhibited at state and county fairs by children. Many families enjoy raising rabbits, particularly show rabbits or raising breeding stock for sale.

Little hands and soft fur; what a wonderful combination!

Chapter 2
Bunny Basics

There's really no such thing as a basic bunny. While there are basic similarities among all types and varieties of rabbits, the vast array of characteristics makes each one an individual. With multiple fur varieties and an entire host of sizes, shapes, and ear types, not to mention the colors and breeds, there truly is a rabbit to suit everyone's preference!

Fur Varieties

One of the first things you might notice about a particular rabbit is its fur. It may be long and fine or short and dense, or it might be somewhere in between. Just as there are numerous varieties and numerous colors of rabbits to choose from, there are also numerous varieties of fur types to consider.

ANGORA

Angora rabbits require consistent grooming and regular care because their unique wool coats are long and dense. Although many rabbit fanciers enjoy the beauty of the Angoras and their impressive wool, owning one requires a definite commitment to regular grooming, otherwise shedding and matting are virtually inevitable. Angora wool is plucked or shaved and can be made into yarn that is used to make clothing.

There are several types of Angora rabbits, each with their own distinct type of coat:

English Angoras are noted for having the finest coats, with particularly long wool on their faces and feet. The wool of the English Angora is silkier than that of the French Angora. Some owners will clip the wool around the rabbit's face to expedite the grooming process.

French Angoras typically require the least amount of grooming due to the presence of additional guard hairs, which help prevent the matting that can occur in the other Angora types. Excessive wool on the face of the French Angora is discouraged. The wool of the French Angora is more coarse than that of the English Angora.

Facing page: A basic bunny? At first glance, perhaps, but in actuality it is a Blue-Eyed White Polish buck. As we will learn in this chapter, there really is no such thing as a basic bunny.

Considerably different from all of the other fur types, Angora wool is beautiful to behold but is rather complicated to maintain. Regular grooming is imperative for all types of Angora rabbits, although French Angoras require somewhat less grooming.

Satin Angoras have coats with less density, but their wool seems to shine with a definite brilliance that the other types of Angoras do not have. This is due to the Satin fur introduced into Angora rabbits, which combined the two popular fur varieties.

Giant Angoras have wool that cannot be plucked because molting does not usually occur, and therefore they need occasional clipping. The Giant Angora's coat has three types of hair: the underwool, the awn fluff, and the awn hair, which is also called the guard hair.

Although they are not Angoras, the American Fuzzy Lop and the Jersey Wooly also have wool coats. These breeds are similar to the French Angora since they have extra guard hairs in their coats and do not require as much grooming as some of the other types.

NORMAL FUR

As its name implies, the term normal fur refers to any breed that does not have a wool, Satin, or Rex coat. Although it is a general catchall phrase for normal fur, there are still many variations in this category. Some coats are known as fly back coats, named for their tendency to immediately return to original position when you've brushed them in the opposite di-

Normal fur, as shown on this Dutch rabbit, is precisely as its name implies: normal. It isn't wool as on the Angora, it lacks the brilliance and shine of the Satin, and it isn't one even length as with the Rex. It's simply fur, and it's found on most breeds of rabbits.

Rex fur is unique from other fur types in that the fur is one consistent length throughout the body without long protruding guard hairs. They are sometimes referred to as Velveteen rabbits.

rection. Other coats are known as roll back because they stay roughed up rather than immediately settling into the normal position when they have been brushed backwards. The density and texture of normal fur depends on the standard for each breed, and there is a good deal of variation between the breeds.

REX

Unlike the long, luxurious coats of the Angoras, Rex fur has an ideal length of only ⅝ inch. Because the guard hairs do not protrude in this type of fur, they are the same length as the undercoat to allow for a consistent length and density

Satin fur is admired for its luster and sheen, which is caused by the transparent effect of the glass-like hair shell displayed in this fur type.

throughout. Rex fur is typically quite thick, and it is also very soft. Some say it feels like velvet. Rex rabbits have been referred to as Velveteen rabbits and are one of the foundation breeds in the recently developed Velveteen Lop breed. The fur is sometimes used for fur coats and hats.

SATIN

Satin fur is renowned for its silky texture and is quite different from the other varieties of fur. The hair shell is transparent and reflective in its appearance, with brilliant coloring that is quite shiny and lustrous. The sheen is unmistakable. The Satin coat is believed to have been a result of a genetic mutation, first observed in the Havana breed in the 1930s and later recognized as its own individual breed, the Satin. There is also a smaller version with the same fur characteristics, the Mini Satin. As we have discussed previously, Satin fur has also been introduced into the Angora, with a breed now known as the Satin Angora.

Shapes

The American Rabbit Breeders Association recognizes five types of rabbits and classifies these types by their unique body shapes.

FULL ARCH

Mainly recognized by the arch that runs from the back of their neck down to their tail, the full arch rabbit is taller than it is wide and is characterized by longer limbs. Breeds included in this type are Belgian Hares, Britannia Petites, Checkered Giants, English Spots, Rhinelanders, and Tans.

Ear Types

Ears that point straight up, parallel to one another, represent the most common type of rabbit ears. The ears on this rabbit are particularly eyecatching due to their coloring.

Lop ears drape down from the head and frame the rabbit's face. The ears are a major consideration in the judging of lop-eared breeds at rabbit shows, with the ears of the English Lop counting for one-third of the rabbit's score when being judged.

What's up? Well, if you own any breed of rabbit that isn't a lop, it's probably your rabbit's ears! Rabbit ears come in two main varieties, the common "up" ears and the less common "lop" ears. Up ears, as the description implies, point straight up over the rabbit's head, while lop ears drape down around the rabbit's face. Lop rabbits are noted to be one of the oldest known types of domestic rabbits and were mentioned in a book from 1854. Of the breeds recognized by the American Rabbit Breeders Association, only five of them have lop ears: the American Fuzzy Lop, the English Lop, the French Lop, the Holland Lop, and the Mini Lop. It should be noted, however, that some types of Angora rabbits have a slightly different ear style. They are not exactly lop, but they are certainly not straight due to the weight of the wool in their ears that pulls the tips downward.

The heavy wool on the English and Giant Angora rabbits cause their ears to droop due to the weight. The ears on the French and Satin Angoras do not have such a significant amount of wool.

This charming Mini Rex is a representative of the compact type of rabbit and is characterized by a short, compact body.

SEMI-ARCH

Less "full" than the full arch is the semi-arch type, where the arch begins behind the shoulders instead of at the back of the neck. This type is also referred to as the mandolin type. Breeds displaying this type are the American, Beveren, English Lop, Flemish Giant, and Giant Chinchilla.

COMPACT

Next to the commercial type, this is the second most common type of rabbit and is characterized by a short, compact body. They are not as popular as the commercial type in terms of meat-producing animals, but they are similar. Breeds in the compact category include the American Fuzzy Lop, English Angora, Standard Chinchilla, Dwarf Hotot, Dutch, Florida White, Havana, Holland Lop, Jersey Wooly, Lilac, Mini Lop, Mini Rex, Netherland Dwarf, Polish, Silver, and Thrianta.

COMMERCIAL

As the name suggests, the commercial types are the best meat rabbits. They have wide, deep, full, and round bodies. This is the most common type of rabbit. Breeds include French Angora, Giant Angora, Satin Angora, Champagne d'Argent, Californian, Cinnamon, American Chinchilla, Creme d'Argent, French Lop, Harlequin, Hotot, New Zealand, Palomino, Rex, American Sable, Satin, Mini Satin Silver Fox, and Silver Marten.

The commercial type is the most common and is characterized by a wide, deep body, as shown on this Satin rabbit.

CYLINDRICAL

The Himalayan is the only breed that exhibits the cylindrical type. The body type is long and slender, like a cylinder.

Sizes

Large and small, rabbits come in sizes to fit all lifestyles. The 47 breeds recognized by the American Rabbit Breeders Association can be divided into four groups based on their ideal mature weight, with most breeds generally ranging from 2 to 11 pounds.

SMALL RABBITS

Generally topping the scales at less than 5 pounds, small-sized rabbit breeds include the following:

American Fuzzy Lop (3 ½ to 4 pounds)
Britannia Petite (up to 2 ½ pounds)
Dutch (3 ½ to 5 ½ pounds)
Dwarf Hotot (2 ½ to 3 pounds)
Florida White (4 to 6 pounds)
Himalayan (2 ½ to 4 ½ pounds)
Holland Lop (3 to 4 pounds)
Jersey Wooly (3 to 3 ½ pounds)
Mini Rex (3 to 4 ½ pounds)
Mini Satin (3 ¼ to 4 ¾ pounds)
Netherland Dwarf (2 to 2 ½ pounds)
Polish (2 ½ to 3 ½ pounds)
Tan (4 to 6 pounds)

Of these breeds, the Britannia Petite, Dwarf Hotot, Holland Lop, Jersey Wooly, Netherland Dwarf, and Polish rabbits are considered to be dwarf breeds, with dwarf-like characteristics and a mature weight of less than 3 pounds. The Netherland Dwarf is considered the most popular, possibly due to the impressive array of colors in the breed. Many rabbit fanciers are drawn to the minute size of the dwarf breeds and enjoy the fact that they can be maintained with less feed and space than larger breeds. However, dwarf litters typically consist of fewer kits than litters of larger breeds, and those planning to raise dwarf rabbits should understand that 25 percent of dwarf kits die due to a lethal genetic combination that occurs in approximately one out of every four kits. This usually happens before the kits are three or four days old.

While they are still small, the American Fuzzy Lop, Dutch, Florida White, Himalayan, Mini Rex, Mini Satin, and Tan rabbits are not dwarf breeds and do not carry the lethal dwarfing gene.

Above: A Mini Lop, like the one pictured here, might sound like a small breed, but it's actually considered a medium-sized rabbit and weighs between 4 ½ and 6 ½ pounds.

Facing page: Dwarf rabbits, such as the ruby-eyed white Netherland Dwarf pictured here, are popular with enthusiasts who enjoy the size of these petite creatures. Small ears and large eyes are characteristic of the dwarf breeds.

MEDIUM RABBITS

Medium-sized rabbits typically range from 6 to 9 pounds and are often raised as show animals. They are occasionally raised for meat, but not as often as the larger breeds due to their smaller size and weight. Medium rabbits include the following breeds:

American Sable (7 to 10 pounds)
Belgian Hare (6 to 9 ½ pounds)
English Angora (5 to 7 ½ pounds)
English Spot (5 to 8 pounds)
French Angora (7 ½ to 10 ½ pounds)
Harlequin (6 ½ to 9 ½ pounds)
Havana (4 ½ to 6 ½ pounds)
Lilac (5 ½ to 8 pounds)
Mini Lop (4 ½ to 6 ½ pounds)
Rex (7 ½ to 10 ½ pounds)
Rhinelander (6 ½ to 10 pounds)
Satin Angora (6 ½ to 9 ½ pounds)
Silver (4 to 7 pounds)
Silver Marten (6 to 9 ½ pounds)
Standard Chinchilla (5 to 7 ½ pounds)
Thrianta (4 ½ to 6 ½ pounds)

LARGE RABBITS

Moving right up the scale are the large-sized rabbits, with mature weights of 9 to 11 pounds. These larger breeds are often raised for meat and for their fur.

American (9 to 12 pounds)

Below: Many large breeds, like the French Lop seen here, display what is known as the commercial body type with a good depth of body that is very round and full.

The largest rabbit breeds are the giants, such as the Flemish Giant. These massive rabbits weigh in at more than 13 pounds and are impressive creatures to behold.

American Chinchilla (9 to 12 pounds)
Beveren (8 to 12 pounds)
Californian (8 to 10 ½ pounds)
Champagne d'Argent (9 to 12 pounds)
Cinnamon (8 ½ to 11 pounds)
Creme d'Argent (8 to 11 pounds)
English Lop (9 pounds and up)
French Lop (10 pounds and up)
Giant Angora (9 ½ pounds and up)
Hotot (9 to 11 pounds)
New Zealand (9 to 12 pounds)
Palomino (9 to 11 pounds)
Satin (8 ½ to 11 pounds)
Silver Fox (9 to 12 pounds)

GIANT RABBITS

Last, but certainly not least, are the giant rabbits that weigh more than 11 pounds, with some weighing in at considerably more. These are popular show rabbits. Breeds in this size range are usually obvious from their names.

Checkered Giant (11 pounds and up)
Flemish Giant (13 pounds and up)
Giant Chinchilla (12 to 16 pounds)

Chapter 3

A Rainbow of Rabbits

We know that rabbits have various types of fur, different varieties of ears, and come in a range of shapes. We also know that they come in a multitude of sizes, from the diminutive Netherland Dwarf to the massive Flemish Giant. Yet there is still another facet of rabbit identification that is much more complicated than anything we have discussed so far: colors.

The Importance of Color

It probably goes without saying that color is a very important characteristic in rabbit breeds. Every breed that is recognized by the American Rabbit Breeders Association has a standard color or colors listed in its standard of perfection. Any rabbit that does not meet the recognized color characteristics of its breed cannot be shown at ARBA shows. However, additional breed color varieties are constantly being developed and fine-tuned, as dedicated breeders work toward the recognition of additional colors and varieties within their specific breeds. For instance, at the 2006 ARBA convention, three new color varieties were approved, including the Chocolate Dwarf Hotot, Sable Point Mini Rex, and Broken Netherland Dwarf. Other colors that were approved as presented but need additional showings before official acceptance and recognition include the Copper Mini Satin, Otter Mini Satin, Broken Havana, Fawn Jersey Wooly, Chocolate Agouti Rex, and Otter Rex. Several additional colors that were presented but failed their presentation include the Brown Beveren, Lilac Havana, and Orange Jersey Wooly. (For more information on the process through which new colors and breeds are approved, please see chapter 6).

The weight of color as a judging factor depends entirely on each breed. In some breeds that are recognized in a multitude of colors, the emphasis on color from a judge's standpoint is minimal. For instance, in a Holland Lop, which is rec-

Facing page: While some rabbits' coats are only one color, some breeds exhibit coats of many colors, such as the tricolored Harlequin (Japanese color) pictured here.

One breed, three colors! The American Fuzzy Lops pictured here exhibit the following colors: (left to right) blue, chocolate, and broken Siamese sable.

ognized in a multitude of color varieties, color and markings count for only 4 points out of a 100-point judging system. This is because it is not the color that makes the Holland Lop what it is. Much more important in the Holland Lop breed are the general body type characteristics, which count for 84 points in the judging system. Similarly, New Zealand rabbits are judged with only 15 points on color and 60 points are attributed to body type. However, some breeds are judged heavily upon their color and markings because these characteristics are so vital to the breed's standard. Consider a Dutch rabbit for a moment. Its distinctive markings are what really differentiates the Dutch rabbit from all of the other rabbit breeds. Because of this, color and markings count for 50 percent of the points when being judged. Other breeds that are judged heavily upon their color or markings include the English Spot, the Silver (like the Dutch, judged 50 percent on color), and the newly recognized Thrianta. Somewhat surprisingly, other breeds, such as the American Chinchilla, are judged primarily on type rather than color, even though there is only one recognized color in the breed.

Because of the importance of color in the show ring (unrecognized colors are not allowed to show), breeders must take care to ensure that the planned mating in their rabbitry meets the standards for the production of offspring in recognized colors. Obviously, this calls for a solid and thorough understanding of rabbit color genetics. Just as important as achieving the ideal standard in colors, markings, and shades of the different varieties is the understanding of how not to produce unwanted colors that are disqualified from showing. This is not always possible, despite efforts on the part of the breeder to minimize these chances. For instance, in the rare American Sable breed, the only recognized color is Sable. Because Sable coloring does not always breed true, it is common for there to be the occasional Californian- or Seal-colored offspring in American Sable litters. Unfortunately, these specimens cannot be shown, but of course they still have good potential as family pets, 4-H rabbits, or possibly as breeding stock for breeders who know how to cross these colors to produce sable. Rabbits in unrecognized colors (or with improper markings) for their breed are known as "sports." Sports have long been used in the production of new rabbit breeds. A good example is the Silver Marten breed, which originated as Chinchilla sports and were admired for their coloring, despite their disqualification from Chinchilla showing.

The importance allotted to color varies according to each breed's standard of perfection. In Holland Lops, such as the one seen here, color counts for only 4 points out of the 100-point judging scale.

Rabbit Color Guide

Attempting to decide which rabbit colors to include in this guide was a lengthy and complicated process. Due to space limitations, it was immediately apparent that we would be unable to thoroughly cover every color variety and shade, and thus we were forced to make some decisions in order to provide a more concise listing of the colors that are encountered in rabbits. Obviously, trying to narrow the list to only 35 colors (from a starting list of more than 100 varieties and shades) was a difficult and time-consuming process, and we hope we have not offended any fanciers of the more unusual colors that we were unable to fit into our listing.

Understanding the Color Guide

Each color listing begins with the name of the color. Some are followed by a listing of possible varieties in shade. Then we have a basic description of the color, along with a photo illustrating the color. This is followed by a listing of breeds in which the color is found. Only breeds where the color is officially recognized by ARBA are listed. This is an important concept to understand when reading the color guide. Many colors appear in breeds other than those listed, but they are either unrecognized or in the process of acceptance for those breeds, and therefore we have not listed them. Finally, we have a listing of the color pattern to which the color belongs to allow for cross-reference within the breed profiles (see chapter 7), as many of the breeds listed in the breed profiles have listings of color patterns. This allows you to cross-reference what colors fit in each pattern.

BLACK

A solid black color overall, uniform, and rich throughout the coat

Found in the following breeds: American Fuzzy Lop, English Angora, French Angora, Satin Angora, Beveren, Britannia Petite, Flemish Giant, Havana, Jersey Wooly, English Lop, French Lop, Holland Lop, Mini Lop, Netherland Dwarf, New Zealand, Polish, Rex, Mini Rex, and Satin

Eye color: Brown

Color pattern: Self

BLUE

Blue or slate; a very rich, deep shade over the entire body

Found in the following breeds: American Fuzzy Lop, English Angora, French Angora, Satin Angora, Beveren, Flemish Giant, Havana, Jersey Wooly, English Lop, French Lop, Holland Lop, Mini Lop, Netherland Dwarf, Polish, Rex, Mini Rex, and Satin

Eye color: Blue-gray

Color pattern: Self

BROKEN

Any coat pattern with a mixture of white and color. In the case of certain breeds (Checkered Giant, Dutch, Hotot, for example) the color pattern must meet specific criteria in order to meet the individual breed's standard of perfection.

Found in the following breeds: American Fuzzy Lop, French Angora, Checkered Giant, Dutch, Dwarf Hotot, English Spot, Hotot, Jersey Wooly, English Lop, French Lop, Holland Lop, Mini Lop, Netherland Dwarf, Polish, Rex, Mini Rex, Rhinelander, and Satin

Eye color: Dependent upon the color variety

CALIFORNIAN

White body with black points (nose, ears, feet, tail)

Found in the following breeds: Californian, Rex, and Satin

Eye color: Pink

Color pattern: Any other variety

CASTOR

Slate blue hair base, then tan, and then dark brown with black hair tips; black outlined ears

Found in the following breeds: Rex and Mini Rex

Eye color: Brown

Color pattern: Agouti

CHESTNUT

Similar to castor coloring, except chestnut is somewhat lighter and more brown than black

Found in the following breeds: American Fuzzy Lop, English Angora, French Angora, Satin Angora, Belgian Hare, Britannia Petite, Jersey Wooly, English Lop, French Lop, Holland Lop, Mini Lop, and Netherland Dwarf

Eye Color: Brown

Color pattern: Agouti

CHINCHILLA

A mixture of three hair colors throughout the body: gray, blue, and black with black hair tips and black-tipped ears. Lighter colored underside.

Found in the following breeds:
American Fuzzy Lop, English Angora, French Angora, Satin Angora, American Chinchilla, Giant Chinchilla, Standard Chinchilla, Jersey Wooly, English Lop, French Lop, Holland Lop, Mini Lop, Netherland Dwarf, Rex, Mini Rex, and Satin

Eye color: Brown   Color pattern: Agouti

CHOCOLATE

Like a milk chocolate candy bar, the chocolate-colored rabbit is a deep shade of brown.

Found in the following breeds: American Fuzzy Lop, English Angora, French Angora, Satin Angora, Havana, Jersey Wooly, English Lop, French Lop, Holland Lop, Mini Lop, Netherland Dwarf, Polish, Rex, Mini Rex, and Satin

Eye color: Brown   Color pattern: Self

COPPER

A mixture of orange, red, and slate blue with black-tipped hairs and ears

Found in the following breeds: English Angora, French Angora, Satin Angora, and Satin

Eye color: Brown

Color Pattern: Agouti

FAWN

Light to dark fawn colored with darker guard hairs. Genetically, it is a diluted orange/tan.

Found in the following breeds: American Fuzzy Lop, English Angora, French Angora, Satin Angora, Flemish Giant, English Lop, French Lop, Holland Lop, Mini Lop, and Netherland Dwarf

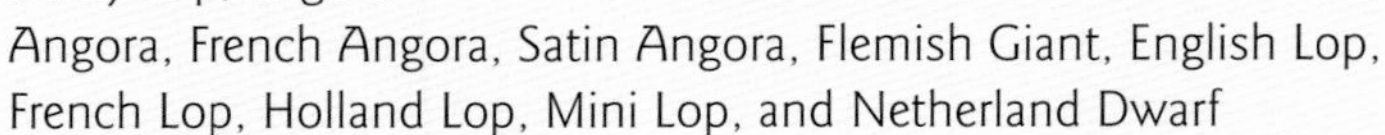

Eye color: Brown

Color pattern: Any other variety

LILAC

A rose-gray color; pinkish lilac. Genetically, it is a diluted chocolate.

Found in the following breeds: American Fuzzy Lop, English Angora, French Angora, Satin Angora, Jersey Wooly, Lilac, English Lop, French Lop, Holland Lop, Mini Lop, Netherland Dwarf, Rex, and Mini Rex

Eye color: Blue-gray

Color pattern: Self

LYNX

White hair base with dark orange center, light orange top, and lilac tips, accompanied by lilac-tipped ears

Found in the following breeds: American Fuzzy Lop, English Angora, French Angora, Satin Angora, English Lop, French Lop, Holland Lop, Mini Lop, Netherland Dwarf, Palomino, Rex, and Mini Rex

Eye color: Blue-gray

Color pattern: Agouti

OPAL

Slate-colored hair base with tan center and blue tips accompanied by blue-tipped ears

Found in the following breeds: American Fuzzy Lop, English Angora, French Angora, Satin Angora, Jersey Wooly, English Lop, French Lop, Holland Lop, Mini Lop, Netherland Dwarf, Rex, and Mini Rex

Eye color: Blue-gray

Color pattern: Agouti

ORANGE

Quite obviously from the name, this is an orange-colored rabbit, and usually quite a bright orange color. Also known as fawn in some breeds.

Found in the following breeds: American Fuzzy Lop, English Lop, French Lop, Holland Lop, Mini Lop, and Netherland Dwarf

Eye color: Brown

Color pattern: Any other variety

OTTER

Comes in four varieties: black, blue, chocolate, and lilac. The body color depends on the variety; the belly is a creamy white color and a shade of tan where the colors meet.

Found in the following breeds: Netherland Dwarf (all four varieties), Britannia Petite (black variety), Holland Lop (all four varieties), Jersey Wooly (black and blue varieties), Rex (black variety), Mini Rex (all four varieties), and Satin (all four varieties)

Eye color: Dependent upon the color variety

Color pattern: Tan

POINTED WHITE

A purely white body accompanied by dark points (ears, tail, feet, nose). Points typically come in four colors: black, blue, chocolate, or lilac.

Found in the following breeds: American Fuzzy Lop (all four colors), English Angora (all four colors), French Angora (all four colors), Satin Angora (black pointed), Himalayan (all four colors), Jersey Wooly (black-and-blue pointed), Holland Lop (all four colors), Mini Lop (all four colors), Netherland Dwarf (all four colors), and Mini Rex (black and blue varieties)

Eye color: Ruby pupil with pink iris Color pattern: Any other variety

RED

Deep, solid, pure red with a somewhat lighter or darker undercoat

Found in the following breeds: English Angora, French Angora, Satin Angora, English Lop, French Lop, Holland Lop, Mini Lop, New Zealand, Rex, Mini Rex, Satin, and Thrianta

Eye color: Brown

Color pattern: Any other variety

SABLE

Deep brown, very dark

Found in the following breeds: American Sable, English Angora, French Angora, Satin Angora, Britannia Petite, English Lop, French Lop, Mini Lop, and Rex

Eye color: Brown

Color pattern: Shaded

SABLE POINT

Creamy body with darker brown points (ears, nose, feet, and tail)

Found in the following breeds: American Fuzzy Lop, English Lop, French Lop, Holland Lop, Mini Lop, Jersey Wooly, Netherland Dwarf, and Mini Rex

Eye color: Brown

Color pattern: Shaded

SEAL

An extremely dark version of Sable coloring (genetically it is homozygous Sable), with a very dark black-and-gray undercolor.

Found in the following breeds: English Angora, French Angora, Satin Angora, English Lop, French Lop, Holland Lop, Mini Lop, Rex, Mini Rex, and Jersey Wooly

Eye color: Brown Color pattern: Shaded

SIAMESE

Light sepia brown on the body, with darker brown along the outline of the rabbit (back, top of tail, ears, and so on)

Found in the following breeds: Satin

Eye color: Brown

Color pattern: Shaded

SIAMESE SABLE

Light sepia brown on the body with darker brown along the outline of the rabbit (back, top of tail, ears, and so on)

Found in the following breeds: American Fuzzy Lop, Jersey Wooly, Holland Lop, and Netherland Dwarf

Eye color: Brown

Color pattern: Shaded

SIAMESE SMOKE PEARL

Overall, a mixture of gray-and-tan hair on the body, with darker gray hair along the outline of the rabbit (back, top of tail, ears, and so on)

Found in the following breeds: American Fuzzy Lop, French Angora, Satin Angora, Jersey Wooly, Holland Lop, Netherland Dwarf, English Lop, French Lop, and Mini Lop

Eye color: Blue-gray

Color pattern: Shaded

SILVER

This color comes in several varieties: black, blue, brown, chocolate, fawn, and lilac. Silver colored throughout the body, accompanied by white hairs.

Found in the following breeds: Champagne d'Argent (black), Creme D'Argent (fawn), English Lop (black, blue, chocolate, and lilac), French Lop (black, blue, chocolate, and lilac), Mini Lop (black, blue, chocolate, and lilac), Silver Fox (black), and Silver (brown, black, and fawn)

Eye color: Dependent upon the color variety

Color pattern: Ticked

TAN

This color comes in several varieties, including black tan, blue tan, chocolate tan, and lilac tan. Body color is dependent upon the overall color of the rabbit (black, blue, chocolate, or lilac), with markings of deep tan or orange.

Found in the following breeds: Netherland Dwarf and Tan

Eye color: Dependent upon the color variety

Color pattern: Tan

SILVER MARTEN

This color comes in four varieties: black, blue, chocolate, and lilac. The overall color depends on the variety with silver markings and hair tips.

Found in the following breeds: Netherland Dwarf, Jersey Wooly, and Silver Marten (each breed comes in all four varieties)

Eye color: Dependent upon the color variety

Color pattern: Tan

TORTOISESHELL

This color comes in four varieties: black, blue, chocolate, and lilac. Reddish-fawn body color with lighter color toward the hair base. Points (nose, ears, feet, tail) are darker with actual color depending upon the variety of Tortoiseshell (i.e., Black Tortoiseshells have black points, Blue Tortoiseshells have blue points, and so on).

Found in the following breeds: American Fuzzy Lop (black and blue varieties only), English Angora (all four varieties), French Angora (all four varieties), Satin Angora (all four varieties), Jersey Wooly (black and blue varieties only), English Lop (all four varieties), French Lop (all four varieties), Holland Lop (all four varieties), Mini Lop (all four varieties), Netherland Dwarf, and Mini Rex

Eye color: Dependent upon the color variety

Color pattern: Shaded

TRI-COLORED

Reminiscent of a calico cat, a tri-colored rabbit features three colors throughout its coat. Often white with black and orange, but tri-colored can be white with blue and fawn.

Found in the following breeds: English Lop, French Lop, Holland Lop, Mini Lop, Mini Rex, and Rhinelander

Eye color: Dependent upon the color variety

WHITE (BLUE-EYED)

Entirely white body color, no colored markings whatsoever, accompanied by blue eyes

Found in the following breeds: American Fuzzy Lop, English Angora, French Angora, Satin Angora, Beveren, Jersey Wooly, English Lop, French Lop, Holland Lop, Mini Lop, Netherland Dwarf, Mini Rex, and Polish

Eye color: Dark blue

Color pattern: Self

WHITE (RUBY-EYED)

Identical to the Blue-Eyed White, with the exception that its eyes are ruby red

Found in the following breeds: American, American Fuzzy Lop, English Angora, French Angora, Giant Angora, Satin Angora, Britannia Petite, Flemish Giant, Florida White, Jersey Wooly, English Lop, French Lop, Holland Lop, Mini Lop, Netherland Dwarf, New Zealand White, Polish, Rex, Mini Rex, Satin, and Mini Satin

Eye color: Ruby-red pupil with pink iris

Color pattern: Self

ADDITIONAL COLORS

There are many other colors that you may encounter in rabbits, including the following:

Cinnamon

Also known as chocolate agouti or amber. As the name indicates, it is a rust-colored red. The color of the breed Cinnamon is also known as cinnamon, but it is a different genetic combination than the cinnamon/chocolate agouti coloring.

Found in the following breeds: English Angora, French Angora, Satin Angora, and Holland Lop

Eye color: Brown

Color pattern: Agouti

Sable Marten

This color comes in four varieties: black, blue, chocolate, and lilac. Dark sepia coloring all along the back, and lighter colored along chest and flanks. Silver colored along points.

Found in the following breeds: Jersey Wooly, Netherland Dwarf, and Silver Marten

Eye color: Brown

Color pattern: Tan

Smoke Pearl Marten

Pearl gray body with lighter gray coloring on the chest and flanks and silver white markings.

Found in the following breeds: Jersey Wooly and Netherland Dwarf

Eye color: Blue-gray

Color pattern: Tan

Squirrel

A coat color with a mixture of blue and gray hairs and blue-tipped ears.

Found in the following breeds: American Fuzzy Lop, English Angora, French Angora, Satin Angora, Jersey Wooly, English Lop, French Lop, Holland Lop, Mini Lop, and Netherland Dwarf

Eye color: Blue-gray

Color pattern: Agouti

Steel

Body color of black, blue, chocolate, or lilac and accompanied by lighter (gold or silver) colored tips.

Found in the following breeds: English Angora (black, blue, chocolate, and lilac varieties), French Angora (black, blue, chocolate, and lilac varieties), Satin Angora (black, blue, chocolate, and lilac varieties), Holland Lop (black, blue, chocolate, and lilac varieties), English Lop (black, blue, chocolate, lilac, sable, and smoke pearl varieties), French Lop (black, blue, chocolate, lilac, sable, and smoke pearl varieties), Mini Lop (black, blue, chocolate, lilac, sable, and smoke pearl varieties), and Netherland Dwarf

Eye color: Dependent upon the color variety

Color pattern: Ticked

Color Genetics

Genetically speaking, rabbit color is controlled by a complicated series of different genes and patterns. The inheritance of color and markings in rabbits is an extensive subject and one we cannot cover in its entirety within the realm of this text. There are books dedicated solely to the explanation of rabbit color genetics so we will give only a brief overview of the basics.

One of the most important points to understand is the difference between phenotype and genotype. The appearance of a rabbit (for instance, a white rabbit with ruby-red eyes) is its phenotype. It's the physical appearance and color that you can see. Alternatively, the genotype is the genetic combination that the rabbit carries, which may or may not be manifested in outward appearance.

For example, two brown Havana rabbits can produce a lilac-colored offspring, but only if both of the brown rabbits carry a recessive dilution gene in their genotype that doesn't appear in their phenotype. Because the dilution gene is recessive, the lilac is hidden, or masked, by the dominant genes that allow for the intense brown coloring to cover up the diluted lilac. It doesn't mean that the recessive gene is

This French Lop and her litter of kits are an excellent example of how color genetics can produce surprises! While some of the kits share their mother's coloring, several of the other kits display additional colors.

This Netherland Dwarf doe with black otter coloring has three kits that all exhibit the same color pattern.

nonexistent, however. It can certainly be passed along to offspring and will certainly manifest itself if both parents pass along a recessive gene to the offspring.

Because genes are inherited in pairs called alleles, it's possible to have three combinations of alleles: two dominant, two recessive, or one of each. If both genes are the same, they are said to be a homozygous pair. If they are not the same, they are said to be heterozygous. Therefore, a rabbit can have a homozygous pair of dominant genes, a homozygous pair of recessive genes, or a heterozygous pair of genes with one dominant and one recessive. In the example of the brown Havana rabbits, both parents carried one dominant and one recessive gene.

It's rarely as simple as two browns producing a lilac, however. Every rabbit has multiple genes that affect its color (phenotypically and genotypically), and it is the combination of these genes that provide us with more than 100 color possibilities in some breeds of rabbits, such as the Holland Lop. These genes control the possibilities of the following patterns: agouti, tan, self, dilute, shaded, ticked, steel, pointed white, and wide band. Consider for a moment the possibilities of all of these genes, each with the potential to be homozygous or heterozygous, and their effect on the color of the rabbit as a whole. It's no surprise that there truly are a rainbow of rabbits!

Chapter 4

Bunny Behavior

Like all animals, rabbits have their own distinct personalities; some are more bold and outgoing and others are more retiring and shy. Some are mischievous and some are gentle. It really depends on the individual; however, some breeds do have reputations for being skittish, while others are noted for being kind and friendly.

Bunnies at Play

Rabbits are typically the most active in the early mornings and evenings, and less energetic during the midday hours. If you want to spend some time enjoying or observing your rabbit, choosing early morning or evening might allow you to see more entertaining antics than you might see otherwise. If you haven't experienced the joy of watching a happy young rabbit enjoying life, then you've missed out on one of life's greatest little pleasures. A happy rabbit enjoying life will race around hyperactively, at extreme speed, bucking and jumping and behaving as if he didn't have a care in the world. (And, at that particular moment, he probably doesn't!)

Communication

In addition to effectively communicating joy, rabbits can also easily communicate other feelings. It is only a matter of learning to identify their meanings in order to understand what your rabbit is saying.

A rabbit that is rubbing its face on its owner's foot is probably seeking attention and is displaying affection. Alternatively, a rabbit that is nipping or turning away is probably saying "leave me alone." Typical displays of aggression can include ears that are pinned back, growling, lunging, and biting. Screaming is an indication of extreme terror. It means the rabbit is thoroughly frightened. Ears bent forward means the rabbit is listening intently, watching attentively, and is very interested in what is going on. A rabbit that is marking its territory (especially a male rabbit or a rabbit in new surroundings) will rub

Facing page: Bath time! This Holland Lop is giving himself a daily grooming. Using his front paws, he rubs his face and ears. Their unique mannerisms make rabbits a delight to observe.

A rabbit with a pleasant personality will be enjoyed by family members of all ages. Rabbits seem to naturally attract admiration, and a sweet disposition only enhances their appeal.

A house rabbit can be a wonderful pet, but it's important to take some simple precautions to ensure the safety and well being of your bunny buddy. By rabbit-proofing a room, you can be certain that your rabbit isn't chewing on something that could cause him harm.

its chin on objects. Teeth grinding is indicative of pain and probably means that the rabbit is injured or sick. On the other hand, a soft chattering of the teeth (reminiscent of the purring of a cat) means that the rabbit is perfectly content and relaxed. The thumping of the hind legs is a signal of danger. This traces back to the wild rabbit's instinct to warn other rabbits in the warren of impending danger. Rabbits may also thump their hind legs when they are anticipating being fed or excited about something. "That's why they call me Thumper," says the baby rabbit on the Disney animated classic *Bambi*.

House Rabbits

While having a house rabbit may be an enjoyable experience for many owners, it's very important to understand the rabbit's natural instinct to chew and dig. Electrical and telephone cords will need to be kept out of reach, as will any items (especially anything made of rubber) that might catch the rabbit's chewing fancy. House plants, especially ones that are toxic to rabbits, should always be kept out of the rabbit's reach. Some owners allow their rabbit to roam in a specific room or two but prefer not to give them free access to the entire house. The rooms that the

A circular spinning toy filled with hay can help keep a bored bunny entertained for hours, as well as keep his tummy filled!

rabbit is allowed to inhabit can be "rabbit-proofed" to make sure that everything is safe and protected.

Rabbits can be trained to use a litter box, and some owners enjoy the company of their house rabbit after it is reliably trained to use the litter box. Training takes patience and time, and not all rabbits are able to be trusted explicitly. Many will make the occasional mistake, and some just aren't able to be trained, despite time and effort on the owner's part.

Proper Handling

The proper handling of your rabbit is very important, with one of the most vital aspects being that you should never pick up a rabbit by its ears. This is unacceptable and could injure the rabbit. A better method of handling your rabbit is to lift him by the scruff of his neck and support him underneath with your other hand. Another possibility is to carefully lift him around his midsection with both hands and then support him against your body. This is the preferred method for owners of show rabbits, as it is felt that lifting a rabbit by the scruff of the neck may possibly damage his fur or eventually loosen the area where you've been lifting him.

Facing page: The proper way to pick up a rabbit is lifting it by the scruff of the neck and immediately supporting the rabbit underneath with your other hand.

Daily examination of your rabbit can help you to detect any signs of illness that you might otherwise miss. A few moments checking your rabbit over will help you recognize anything out of the ordinary.

Recognizing Illness

Another important aspect of bunny behavior is understanding how rabbits act when they are sick or suffering from pain. Daily observation of your rabbit and its habits will help you recognize when something is wrong. Keep your eyes open for any signs of possible illness, such as lethargy, loose stools, limping, or teeth grinding. You will also want to be aware of any abnormal swelling on the rabbit's body, as well as any change in respiration, any posture changes, and nasal discharge. Lack of hunger can be another signal of illness. If you suspect that your rabbit may be ill, immediately contact your veterinarian.

Rabbit Facts

Rabbits are not rodents. They belong to the order of *Lagomorpha* and the family of *Leporidae*. The major distinguishing factor between Lagamorphs and rodents is that the Lagomorphs have a second set of incisors that rodents do not have.

Rabbits cannot vomit.

The Easter Bunny tradition evolved because of the festival of the Teutonic goddess of spring and fertility, known as Eastre. One of the traditions of her festival was the Easter Rabbit, and this festival eventually became connected to the Christian holiday, Easter.

The largest litter of rabbits ever recorded was 24 kits.

The eyes of a rabbit are very sensitive to light. They have a range of vision of 190 degrees per eye.

The average lifespan of a domestic rabbit is approximately 5 years, with a possible life span of 15 years. Wild rabbits rarely survive past the age of 2 years old.

The average normal body temperature for a rabbit is 101 to 104 degrees Fahrenheit. The average respiratory rate is 30 to 60 breaths per minute. The average heart rate ranges from 130 to 325 beats per minute, although this figure is dependent on the individual rabbit and its activity level.

The average length of gestation in rabbits is 28 to 31 days.

Chapter 5

"Kindling" Spirits

Is raising rabbits in your future plans? Do you dream of starting up a rabbitry and producing top-quality examples of your favorite breed? Perhaps you're drawn to the idea of promoting a rare breed, or you would like to produce that ideally marked English Spot that precisely meets the standard of perfection. In any case, sooner or later, you're going to be dealing with pregnancy and kindling. In this chapter we will go over the basics of these important topics, although we do recommend that you thoroughly research and study these topics or read some of the excellent books that specialize in rabbit breeding. The birth of kits is one of the most important things that will happen in your rabbitry, and you want to be a fully prepared and knowledgeable breeder.

The Pregnant Doe

The optimal age for breeding a doe is when she is between the ages of 6 months and 5 years. Fertility decreases past the age of 5, as does the health and condition to promote healthy pregnancies and offspring. Unlike most animals, rabbits are induced ovulators, and do not have regular heat cycles in the way that many other animals do. Induced ovulation means that ovulation occurs after mating. Because of the increased odds of fertilization due to the increased ovulation, 85 percent of rabbit breedings result in pregnancy.

Pregnancy diagnosis in rabbits can be somewhat difficult to ascertain, but the most common and accurate method of diagnosis is abdominal palpation at 11 to 14 days of pregnancy. The growing kits can be felt, each the size of a small grape, without undue risk to the doe or her kits. An alternative method of pregnancy detection, which involves a test breeding to see if the doe is receptive, is less reliable and also carries increased risk. Not all pregnant does will refuse the buck, and if bred subsequently, it is possible for a second pregnancy to interfere with the initial

Facing page: Undoubtedly one of the most adorable baby animals, a baby rabbit (kit) doesn't open its eyes until approximately 12 days old.

Does line their nests with fur to provide a warm layer around their newborn kits. Although they are not typically pink as shown here (just because it's cute!), these fur-lined nests keep the kits comfy during the first days of life.

pregnancy. For these reasons, it is generally recommended to rely upon the abdominal palpation method to detect pregnancy.

The gestation period for rabbits is between 28 and 31 days, with a couple of days of variation on either side. It has been noted that the American breed of rabbit typically has a gestation of 31 to 33 days. Baby rabbits (kits) delivered prior to 26 days of gestation have a significantly reduced chance of viability.

At 26 to 28 days of pregnancy, a nest box filled with straw or hay, perhaps lined with wood shavings (not cedar) underneath the straw should be provided in the doe's cage. Shredded paper is also another option for bedding. Most does will use the nest box to begin preparations for the nest for her impending young and will line the nest with fur pulled from her body.

Raising Kits

Birth typically occurs during the night or early morning hours. Dystocia (difficult birth) is uncommon in rabbits, and the doe will likely deliver her young without any difficulty. It is important to note that the doe's surroundings should be kept quiet and calm, as stress and anxiety can lead to the doe adopting a cannibalistic attitude toward her kits. It is also important to verify that she has delivered the kits inside the nest box. It is not uncommon for a doe to deliver her kits in various places throughout her cage, and there is the dangerous possibility that the kits may slip through the cage floor. You will want to vigilantly watch for this potential problem, and promptly move any stray kits back into the nest box where they will be safe.

Kits are born blind, deaf, and mostly hairless. Does have between 8 and 10 mammary glands, which means that most are amply able to provide for the needs of the average litter. In the case of does with very large litters, some breeders will transfer a few extra kits to another doe with a smaller litter. The kits will live in the nest box for approximately two weeks, and the doe will visit the nest once or twice a day to feed her young. After about 10 days, the kits' eyes begin to open and they will start to venture outside of their nest box, begin to eat solid food, and become more

One little happy family! This Netherland Dwarf and her darling kits are the happy sight that all breeders hope for when they begin raising rabbits.

All rabbits should have continual access to fresh water, but this is especially vital for nursing does who need fresh water at all times in order to avoid dehydration.

curious and interested in the world. If after 12 days, the kits' eyes haven't opened, it is recommended that the owner gently help open their eyes. It is also important to vigilantly watch for eye infections, as they are quite common in newborn rabbits.

When the kits are 4 to 6 weeks old, the does will begin to gradually wean their kits over the course of a week or two. By 6 to 8 weeks, the litter should be fully weaned and moved to new cages. At this time, many breeders opt to separate the litter by sex, although it is possible to wait until the kits are 3 months old to do this.

Rabbit breeders will often breed their doe for her next litter when her kits are approximately 6 weeks old. She can finish caring for her current litter until they are 8 weeks old, and then have 2 weeks of rest before the next litter arrives. Four litters per year is considered optimum for many does.

A phenomenon that occurs in the dwarf rabbit breeds is the occasional production of a super-tiny baby rabbit in an otherwise normal-sized litter. These undersized kits are known as peanuts, and the tiny size is due to the fact that these kits inherited two copies of the dwarf gene, making them "double dwarfs." This is nearly always a lethal condition, and most peanut kits die before the age of 10 days. This double dwarf problem only occurs in dwarf breeds, such as the Netherland Dwarf and the Holland Lop.

Although these kits are living happily together in the same cage, it's important to separate the males and females of the litter before they are 3 months old.

2kg
CAPACITY 4lb / 2kg
SENSITIVITY 1oz / 10g
LEGAL FOR USE IN TRADE
1kg

Chapter 6

Show Me the Bunny

Sooner or later, it happens. You're walking by your rabbit's cage one morning when the thought crosses your mind that Fluffy really is a good-looking doe. In fact, she's quite remarkable. You slowly walk around, admiring her from each side, when you get the idea that it might be fun to take her to a show. The breeder you purchased her from had mentioned that she would make a nice show rabbit and now just might be the time to give it a try. But where do you start? Where do you find a show? How do you learn about the proper protocol for showing? While these questions may seem daunting to a first-time rabbit show attendee, you will probably find that you learn the ropes sooner than you might expect, especially when you are aided by the friendly and helpful folks you will meet at the rabbit shows. But for now, let's get our feet wet with a little overview of rabbit showing.

The World of Rabbit Shows

At the 83rd annual American Rabbit Breeders Association convention in October 2006, more than 16,000 rabbits were shown in the Open and Youth divisions, with all 47 recognized breeds represented. The Mini Rex breed had the most representatives, with 1,465 examples shown in that breed. Other popular breeds included the Netherland Dwarf (more than 1,000 shown), the Holland Lop (872 shown), and the Mini Lop (nearly 600 shown). On the contrary, only 6 Satin Angora rabbits, 7 Giant Angora rabbits, and 11 American rabbits were shown. The Best in Show in the Open category was a Dutch rabbit, who was awarded this auspicious title after winning its Best of Breed award, as well as the Best of Group II award.

Facing page: Weight is an important factor at shows, and rabbits must fall within the required weight limitations for their breed. If they do not, they will be disqualified from showing.

Rabbit
Judging
9:00am
Friday

The moment that all rabbit exhibitors plan and prepare for! "Rabbit Judging at 9:00 a.m." Oh, the nervousness and anticipation of the big moment!

What is it that draws thousands of exhibitors to an event with more than 16,000 rabbits in tow? Obviously, there is great enthusiasm for rabbit showing in the United States! Whether it's showing locally at 4-H events, at the county fair, or perhaps at larger ARBA-sanctioned shows, rabbit breeders and enthusiasts have long enjoyed the allure of showing and striven to achieve show quality in their rabbitries. Rabbit shows are a wonderful place to make new contacts and put you face to face with other enthusiasts and breeders. If you're still trying to decide which breed is right for you, a rabbit show is a wonderful place to see a multitude of breeds at one time and help you better visualize the size and type of each breed. Rabbit shows can be extremely educational, as you watch the judging unfold and each exhibit compared to the ideal standard. They are fabulous places to make new friends who are also enthusiastic about rabbits and offer a splendid opportunity for rabbit owners to display their stock to other breeders. Many rabbits are bought and sold at every rabbit show as people seek to improve their rabbitry, introduce different colors or breeds, or simply increase the overall quality of their rabbits. For breeders, a rabbit show is an ideal place to have the results of their breeding program evaluated against other specimens. With all of these attributes, is there any wonder that rabbit shows are immensely popular gatherings?

At ARBA-sanctioned shows, rabbits are governed by rules that you might not find at a local rabbit show or 4-H event. Each breed is judged against the individual breed's standard of perfection in separate classes based upon age or variety. Typically, classes are divided into junior rabbits (those less than six months of age) and senior rabbits (those more than six months old). The larger breeds (those with an ideal senior weight of more than 9 pounds) also have a third division, known as 6/8s, or intermediates, for 6- to 8-month-old rabbits. ARBA recognizes these class divisions as four-class rabbits and six-class rabbits. (See the sidebar on page 84 for which breeds are shown in which division.) Some breeds are also divided by color; there may be separate classes for broken or solid coloring. The first prize winners of each class within a breed are brought together for the judge to choose a Best of Breed winner, and then a Best Opposite Sex winner is chosen from the remaining prize winners of the opposite sex than the Best of Breed winner.

As noted previously, the annual ARBA convention is the largest rabbit show of them all and attracts the largest numbers of exhibits and exhibitors. The convention location rotates around the United States, although it has been noted that the conventions held in the midsection of the United States typically have larger attendance than conventions on either coast, due to the simple logistics of driving distance. For many rabbit breeders, the ARBA convention is the highlight of their year and is a chance to take part in something that has been an annual event since 1924. The conventions typically run for several days, from a Friday through the next Wednesday, and contain an exciting array of events, including specialty club meetings and banquets, ARBA's annual general meeting, social events, and rabbit judging. The ARBA convention is a must-attend for any rabbit enthusiast. It allows rabbit owners to see every recognized breed in one location and gives them the chance to talk with and learn from longtime breeders.

Showing Your Rabbit

If ARBA-sanctioned shows are your aim, it is highly recommended that your first step be to join ARBA. In order for your rabbit's points to count toward awards, you need to be an ARBA member. In addition to the ability to earn points at sanctioned shows, you will also receive the bi-monthly magazine *Domestic Rabbits*, a copy of the ARBA

Members of 4-H often show their rabbits at state and county fairs. This Harlequin rabbit earned third place in his division.

Yearbook, and a copy of the ARBA *Official Guide Book to Raising Better Rabbits and Cavies*. It is an excellent value for your membership fee. It is not required that you join ARBA in order to show at a sanctioned show, but it is recommended.

It's also recommended that you join any breed specialty clubs for the breed or breeds you are planning to show. Supporting these organizations helps the breed and can help establish you as a serious breeder. It's also a good idea to join any local or state rabbit clubs, as they will likely be hosting many of the events that you will attend. Again, supporting these groups helps to ensure that rabbit shows continue to flourish in your area.

Another important step is to purchase the *Standard of Perfection* book from ARBA. This book lists the ideal standards of the perfect rabbit in each breed. It also lists other important criteria, such as ideal weights and accepted colors, in addition to the description of the rabbit's ideal physical characteristics. You will want to familiarize yourself with the *Standard of Perfection* so you can choose the best rabbits to show.

Above: Off to the show! These rabbits are traveling in special travel cages. Once at the show, they will be settled into the show cages and supplied with water and food.

Facing page: All rabbits shown at ARBA-sanctioned shows must have a tattoo in their left ear. The tattoo is a combination of numbers and letters to help keep track of all entries at a rabbit show.

NED

In order to show at an ARBA-sanctioned show, all rabbits must have a tattoo in their left ear. The sequence of numbers/letters is part of the way entries are identified at the show. After all, at a show with thousands of rabbits, the double-checking of identity can be a very important thing. Breeders decide upon the arrangement of numbers and letters for the tattooing in their rabbitry. This allows the breeder to keep track of their rabbits for identification and record keeping. At a large rabbitry with many rabbits, the tattoo can be an important part of keeping track of "who is who" within the rabbitry.

It is also important to fully understand the list of potential disqualifications, as an animal possessing one of the items will be disqualified from competition. Common disqualifications (DQs) include: a tattoo in the wrong ear (must always be in the left ear), malocclusion of the teeth, toenails that are the wrong color, a rabbit exhibiting signs of illness, or a rabbit exhibited in the wrong class (double-check what you are entering and where). For instance, American Blues, Japanese Harlequins, and Lilacs can all be disqualified for having white toenails. Jersey Wooly rabbits can be disqualified for having ears that are more than 3 inches long, and Britannia Petites can be disqualified for having a "bulldog head."

Once you've decided on a show to enter, you will want to fill out your pre-entry forms and mail them to the show secretary. Double-check to make sure that you have filled out all paperwork correctly and included any necessary entry fees. Then it's time to begin preparing your rabbit for the big day!

On the day of the show, you will obviously want to have your rabbit in tip-top show condition, fully groomed, and ready to show. Pay close attention once you've arrived at the show to make sure that you are aware of how the show schedule is flowing and how much time you have before your class. It goes without saying that you will want to be on time for your class. It's always a nerve-wracking time for the exhibitor while the judging is underway, but carefully listening to any comments the judge may have is educational and enlightening. If all goes well, you just might come home with the first place prize!

Help! My Rabbit Has Three Legs!

To the novice rabbit owner, it can be a rather unsettling event when you hear that a rabbit has an abnormal number of legs. You are browsing the website for a rabbitry that raises the breed you are interested in, admiring their stock for sale, when suddenly, you notice a rabbit for sale with the following information. "Sable point buck, nice head, three legs."

Three legs? Your eyes widen and you begin to wonder what kind of second-rate rabbitry this might be. Scrolling farther down the page, the news gets worse. There is a chestnut agouti doe with two legs and a lynx buck with only one leg! You start scratching your head, wondering what sort of cannibalist do they have in their rabbitry, when suddenly you realize that perhaps legs might mean something else with rabbits.

The good news is, it does! Rabbits shown at ARBA-sanctioned shows have

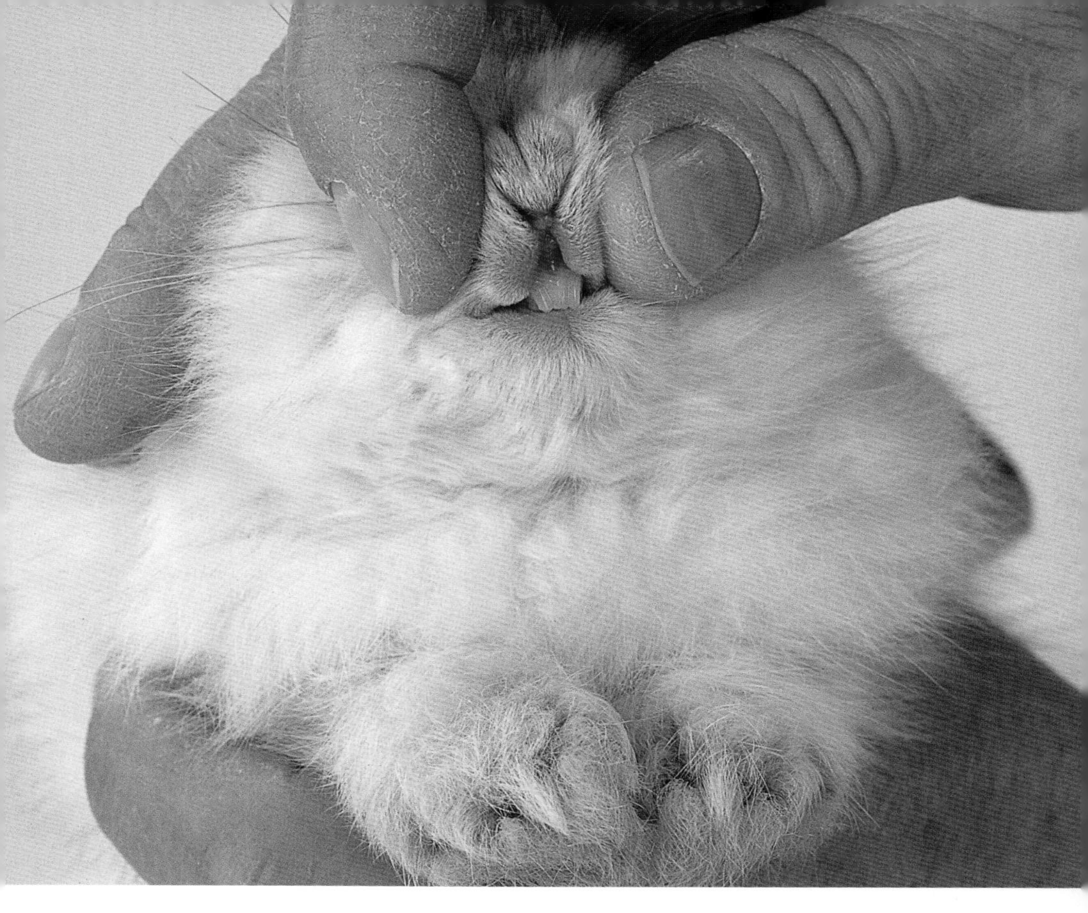

There are several faults that can cause a rabbit to be disqualified from a show. Malocclusion (misaligned teeth) is one type of disqualification. Check your rabbit's teeth before you head off to a show.

the opportunity to earn "legs." There are a variety of ways to earn legs, based on a rabbit's winnings and the number of entries it was shown against. For example, placing first in a class of five or more entries with three or more exhibitors qualifies the rabbit for a leg. Winning a category, such as Best of Breed or Best of Group, with the same regulations also qualifies the rabbit for a leg. While some additional rules do apply, such as showing under at least two different ARBA-licensed judges, generally speaking, when a rabbit has collected three legs the owner can apply for a Grand Champion Certificate from ARBA. Grand Championships can be recorded on pedigrees and is a positive attribute that breeders can mention when selling the rabbit's offspring. In short, finding a rabbit with three legs is a very good thing!

Four- and Six-Class Rabbits

Rabbit breeds whose ideal weight (senior) is less than 9 pounds are shown as a Four Class Breed. Rabbit breeds whose ideal weight (senior) is more than 9 pounds are shown as Six-Class Breeds. The classes in the Four-Class Breeds are Senior Buck, Senior Doe, Junior Buck, and Junior Doe. In the Six-Class Breeds, the classes are Senior Buck, Senior Doe, Intermediate Buck, Intermediate Doe, Junior Buck, and Junior Doe. The breeds fall into these groupings as follows:

FOUR-CLASS BREEDS

American Fuzzy Lop
American Sable
English Angora
French Angora
Satin Angora
Belgian Hare
Britannia Petite
Standard Chinchilla
Dutch
Dwarf Hotot
English Spot
Florida White
Harlequin
Havana
Himalayan
Jersey Wooly
Lilac
Holland Lop
Mini Lop
Netherland Dwarf
Polish
Rex
Mini Rex
Rhinelander
Mini Satin
Silver
Silver Marten
Tan
Thrianta

SIX-CLASS BREEDS

American
Giant Angora
Beveren
Californian
Champagne d'Argent
Checkered Giant
American Chinchilla
Giant Chinchilla
Cinnamon
Creme d'Argent
Flemish Giant
Hotot
English Lop
French Lop
New Zealand
Palomino
Satin

From Hare to There: The Road to Becoming a Sanctioned Breed

Perhaps you've wondered about the process through which the 47 ARBA-recognized rabbit breeds have achieved their official recognition. Take the Thrianta, which was officially recognized by ARBA in October 2005. How did the Thrianta achieve its recognition? The process is a long and complicated one and requires a great deal of commitment and hard work. It is not, as we will see, a process for the faint of heart.

The first step toward the recognition of a new breed or variety is the issuance of a certificate of development (COD) from ARBA. Longtime ARBA members (those with a minimum of five years of membership) can request a COD after submitting a request to ARBA, complete with a sample written standard for the new breed and an application fee. If ARBA approves the request, the breeder receives the COD and is allowed to begin the process of achieving breed sanctioning.

In order to become a sanctioned breed, the breed needs to be presented three times at an ARBA National Convention and be successfully approved each time. However, before the presentations can occur, the COD holder must have held the certificate for three or more years. Once the first presentation has been made, the breeder has five years to achieve three successful presentations. If two consecutive presentations fail, then the breeder is considered to have failed the certificate and the next COD holder in line is allowed to proceed with their work on the breed. On the other hand, after the third successful presentation in a five-year period, the breed is considered to have achieved full recognition and becomes eligible for showing in the following year.

There are also regulations on the animals shown at the presentations. The first presentation must be comprised of two pairs, a senior pair and a junior pair, with the junior pair being the senior pair's offspring. Subsequent presentations must be comprised of three pairs, with one of the three pairs being from an earlier presentation.

The Lionhead breed was exhibited for its first presentation in October 2005 at the ARBA National Convention and was successful. The second presentation occurred in October 2006 and was a failed presentation due to a disqualifying white spot on one of the presentation does. While this is a minor setback for enthusiasts of the breed, if the 2007 presentation is successful, then it is hoped that 2008 will hold the third and final successful presentation that the breed needs for official recognition. However, if the 2007 presentation fails, then the next COD holder in line will be given a chance to bring their presentations (assuming that they have held their COD for a minimum of three years).

Showing is an exciting experience and a recreation that many families enjoy with their rabbits. Involving the entire family makes going to shows even more special.

Another good example is the Velveteen Lop breed, which has faced several setbacks, including failed presentations, COD holders who have withdrawn their certificates, and failures to show. Each time the COD passes to the next holder, the process begins again and the breed (despite more than a decade of attempts) awaits official sanctioning.

Obviously, rabbits of any breed or mix can be shown at rabbit shows, but to compete at ARBA-sanctioned shows, the breeds must be sanctioned by ARBA. However, specialty clubs occasionally host shows for specific breeds, whether or not they are ARBA sanctioned. An excellent example of this is the North American Lionhead Rabbit Club, which held a National Exhibition show in 2006, and attracted more than 600 Lionhead rabbits, despite the fact that the breed is not yet officially sanctioned.

These French Lop rabbits are relaxing in their cage.

Chapter 7

Breed Profiles

How to Use the Breed Profiles

There's a good chance that you have already flipped through the breed profiles, looked over the photographs, and perhaps have a preliminary idea of which breeds catch your fancy. Now you are ready to sit down and begin reading through each breed to compare them. Or perhaps you're at a rabbit show and have just seen your first Harlequin or Creme d'Argent and are thinking, "That's a beautiful breed, I should find out more about it." Or maybe you are wandering through the county fair and just passed by a cage with a gorgeous rabbit inside, but no breed information is listed on the cage. Your curiosity is piqued, but what to do? How do you find out more? The answer is easy; you simply reach into your pocket for this book and flip to its section on breed profiles. Undoubtedly, the breed profiles are the most helpful if you are acquainted with what all of the different categories mean. Let's take a few moments to go over each category so that you'll know how to use them most effectively.

Breed name: You probably figured this one out by yourself! When we say "breed name," we are referring to the officially recognized version of the breed name as listed in the American Rabbit Breeders Association *Official Guide Book to Raising Better Rabbits and Cavies*. In some instances, the breed name has undergone changes over the course of history, especially in the case of breeds that have been imported to the United States from foreign countries (as opposed to breeds developed in the United States). In some cases, the original name recognized by ARBA has been changed over time. One example is the Hotot rabbit, which was

The American Rabbit Breeders Association recognizes 47 breeds, including the Dutch, as seen here. Additional breeds are in the process of being sanctioned by ARBA, and it's possible that the list will continue to grow.

originally known as the Blanc de Hotot, and the name still appears in this variety in some sources. However, we have listed the names as they currently appear in ARBA literature.

Also known as: This is the breed's slogan. It's a catchy phrase that showcases and highlights the breed's outstanding characteristic.

Breed description: This is a brief overview of the history of the breed: how it was developed, where it originated, which breeds were instrumental in its development, and so on. Important physical characteristics of body type, color, fur, or markings are also noted. The breed's history with ARBA is noted in some cases.

Size: As discussed in chapter 2, rabbits come in a variety of sizes, from small to giant. The general size of the breed is noted based on its range of acceptable weight.

Shape/type: As discussed in chapter 2, rabbit breeds are found in a variety of shapes. ARBA recognizes five body types. The shape required by ARBA for each breed is noted.

Colors: Some breeds are only recognized in one standard color, and if so, it has been noted, along with a brief description of the color. However, most breeds come in more than one color, or in many cases, an entire rainbow of colors. For these, there are listings of the specific colors. For breeds where the listing is particularly extensive, we have restricted our description to a listing of the color patterns that are accepted in the breed. Please cross reference to chapter 3 for specific colors that appear in these breeds.

Weight: Nearly every breed has separate weight guidelines for bucks and does, as well as separate weight guidelines for registration or ideal. Generally, the ideal weight is a much more specific figure than the more generalized registration weight figure. Each of these figures is noted for each breed, as officially recognized by ARBA's standard of perfection for each breed.

ALBC status: The American Livestock Breed's Conservancy (ALBC) is a non-profit organization founded in 1977 to help preserve and promote rare and endangered livestock breeds. The ALBC currently highlights several ARBA-recognized rabbit breeds that have global populations of fewer than 2,000 animals, and in some cases, fewer than 500 animals. These breeds are on the ALBC's critical, threatened, watch, and study lists. Obviously, when a rabbit breed has fewer than 2,000 specimens worldwide, it is in a much greater risk of extinction than a rabbit breed that has 1,200 specimens at a single show. Thus, the breeds that are on the ALBC list are noted here. If you are looking to raise rare breed rabbits or help preserve a breed that is nearing extinction, you will definitely want to make note of the breeds on the ALBC lists.

Breed Specialty Club Website: In addition to the excellent information available at the American Rabbit Breeders Association website (www.arba.net), there are also numerous websites on the internet that are maintained by breed specialty clubs. These websites can be a wonderful resource for anyone seeking to learn more about a certain breed. Many of these websites feature photographs, specialty show information, breed history, news, and links to breeders and enthusiasts who can assist you in learning more about their chosen breed. There is truly a wealth of information at your fingertips while browsing these club websites.

What are you waiting for? The world of rabbits awaits!

American

Also known as: Established as a Classic

Breed description: Although very popular in the 1920s, the American breed of rabbit has become rare today. It was developed in Pasadena, California, in the early 1900s by Lewis H. Salisbury. Though he would never disclose the breeds through which he developed the American rabbit, its mandolin shape and blue color is believed to have come from the blue breeds common in America at the time (i.e., Blue Flemish Giants and Blue Beveren). The breed's original name was the German Blue, but it was renamed after World War II. The Blue varieties of the American rabbit are noted for their dark, rich blue coloring. The less common White variety of the American is believed to have been developed from throwback "sports" of the American Blues, the occasional white rabbit produced from mating blue rabbits. These Whites were bred and possibly crossed with White Flemish Giants to achieve the white coloring. Nearly extinct today, American rabbits are still produced by a few dedicated breeders who hope to maintain this breed and Lewis H. Salisbury's vision. In 2006, a herd of American Whites was discovered in Canada. The rabbits from this breeding program had been incorrectly referred to as New Zealand Whites for approximately 80 years, when in reality they were the rare American Whites.

Size: Large

Colors: Blue, White

Shape/type: Semi-arch

Weight:
Ideal buck: 10 pounds
Ideal doe: 11 pounds
Registration buck: 9 to 11 pounds
Registration doe: 10 to 12 pounds

ALBC status: Critical (estimated global population of fewer than 500)

Breed specialty club website:
www.rabbitgeek.com/abwrc.html

American Fuzzy Lop

Also known as: Head of the Fancy

Breed description: The American Fuzzy Lop has a unique and interesting history. The breed was established through the efforts of Holland Lop breeders trying to introduce broken coloring into their solid-colored Holland Lops. To achieve this, English Spots were crossed with the Holland Lops. While the initial result of broken-colored Holland Lops was reached, the flyback coats of the English Spots began appearing in the Holland Lops. The breed standard for Holland Lops calls for a rollback coat. To counteract the appearance of the flyback coats, French Angoras were crossed on the Holland Lops in an attempt to reintroduce higher quality fur. At this point it so happened that the occasional Holland Lop was born with Angora wool. Some breeders noted the excellent possibilities for a lop-eared rabbit with wool, especially in the small size, and began breeding for these characteristics. In the 1980s, breeder Patty Green-Karl began presenting her American Fuzzy Lops at the ARBA conventions, and in 1988 the breed was officially recognized. The American Fuzzy Lop comes in 19 recognized colors, including broken and solid colors.

Size: Small

Colors: Chinchilla, Chestnut, Lynx, Opal, Squirrel, Pointed White, Black, Blue, Blue-Eyed White, Chocolate, Lilac, Ruby-Eyed White, Sable Point, Siamese Sable, Siamese Smoke Pearl, Tortoiseshell, Fawn, Orange, and Broken

Shape/type: Compact

Weight:
Ideal buck: 3 ½ pounds
Ideal doe: 3 ¾ pounds
Registration buck: not more than 4 pounds
Registration doe: not more than 4 pounds

ALBC status: None

Breed specialty club website:
http://users.connections.net/fuzzylop/

American Sable

Breed description: Produced from the "sports" (throwbacks) of the Chinchilla rabbit breed in France, the original Sables were considered an undesirable color in the Chinchilla breed. However, some breeders in the early 1900s began crossing these sports and found that they typically bred true with their sable coloring. The first club for these rabbits (British Sable Rabbit Club) was established in 1927 and recognized two colors: the Siamese and the Marten. Meanwhile in the United States, Chinchilla rabbit breeders were experiencing the same phenomenon; the occasional production of a Sable-colored rabbit. The first were noted in 1924 and the American Sable Rabbit Association was formed in 1929. However, despite interest in the breed in the early years, it was nearly extinct in the United States by the early 1980s and was going to be removed as an ARBA-recognized breed, unless a certain number of representatives were shown at the ARBA conventions in 1982 and 1983. It was only through the efforts of dedicated enthusiasts that the breed endured, with the formation of the American Sable Rabbit Club in 1982 and the subsequent recovery of the breed.

Size: Medium

Shape/type: Commercial

Colors: Sable

Weight:
Ideal buck: 8 pounds
Ideal doe: 9 pounds
Registration buck: 7 to 9 pounds
Registration doe: 8 to 10 pounds

ALBC status: Study (There are approximately 500 to 800 American Sable rabbits in the United States today.)

Breed specialty club website:
www.americansablerabbit.com

English Angora

Also known as: The Bunny with a Bonus

Breed description: It is believed that the ancestors of the English Angora originated in Turkey. This unique, silky wooled rabbit has a long history, with documentation of Angora rabbits dating back to the 1700s. It spread through Europe and was brought to the United States during the first half of the nineteenth century. The official ARBA guide book says that the English Angora should give the "appearance of a round ball of fluff." The breed is characterized by a wooly face, as well as wooled legs, feet, and tail.

Size: Medium

Shape/type: Compact

Colors: Any color in the following varieties: Agouti, Pointed White, Self, Shaded, Ticked, Wide Band

Weight:
Ideal buck: 6 pounds
Ideal doe: 6 ½ pounds
Registration buck: 5 to 7 pounds
Registration doe: 5 to 7 ½ pounds

ALBC status: None

Breed specialty club website:
www.nationalangorarabbitbreeders.com

French Angora

Also known as: The Bunny with a Bonus

Breed description: Historically speaking, the French Angora is similar to the English Angora. However, the French Angora's development as a commercial rabbit in France occurred earlier than the English Angora's development as a show animal in England. As the two varieties developed individually, they developed their own characteristics. The French Angora is noted for having a more coarse coat than the English Angora and is also somewhat larger. The head, ears, and legs are not as heavily wooled as the English Angora.

Size: Medium

Shape/type: Commercial

Colors: All colors in the following groups: Agouti, Broken, Pointed White, Self, Shaded, Ticked, Wide Band

Weight:
Ideal buck: 8 ½ pounds
Ideal doe: 8 ½ pounds
Registration buck: 7 ½ to 10 ½ pounds
Registration doe: 7 ½ to 10 ½ pounds

ALBC status: None

Breed specialty club website:
www.nationalangorarabbitbreeders.com

Giant Angora

Also known as: The Bunny with a Bonus

Breed description: Specifically developed in an attempt to create a larger rabbit with Angora wool, the Giant Angora has a background of German Angora (larger Angoras bred in Germany), French Lop, and Flemish Giant. The breed was perfected during the 1980s by Massachusetts rabbit breeder Louise Walsh. ARBA accepted the breed in 1988. Unlike the French and English Angoras that are recognized in a multitude of colors, the Giant Angora is only recognized in the Ruby-Eyed White variety.

Size: Large

Shape/type: Commercial

Colors: Ruby-Eyed White

Weight:
Registration bucks: 9 ½ pounds and up
Registration does: 10 pounds and up

ALBC status: None

Breed specialty club website:
www.nationalangorarabbitbreeders.com

Satin Angora

Also known as: The Bunny with a Bonus

Breed description: With the interest in both satin fur and Angora wool, it seems only natural that a breeder would undertake to produce a rabbit with a combination of the two characteristics and create a rabbit with long wool and a satin sheen. A couple of breeders attempted to combine the two breeds in the 1930s but did not achieve much success. A Canadian rabbit breeder began working toward the Satin Angora in the 1980s using French Angoras and Satin rabbits. The Satin Angoras received ARBA approval in 1987. While the French Angora is recognized in both solid and broken color patterns, the Satin Angora is currently only recognized in solid colors. However, a certificate of development exists for the broken variety, and we may see broken Satin Angoras in the show ring in future years.

Size: Medium

Shape/type: Commercial

Colors: Agouti, Pointed White, Self, Shaded, Ticked, Wide Band

Weight:
Ideal buck: 8 pounds
Ideal doe: 8 pounds
Registration buck: 6 ½ to 9 ½ pounds
Registration doe: 6 ½ to 9 ½ pounds

ALBC status: None

Breed specialty club website:
www.nationalangorarabbitbreeders.com

Belgian Hare

Also known as: King of the Fancy

Breed description: Once a hugely popular breed, the Belgian Hare is credited with being the foundation for the rabbit industry in the United States. Although they were not the first domestic rabbits in this country, they were among the first breeds imported, with the first Belgian Hares coming to the United States in 1888. Prior to this, they were popular in England and Belgium. It is believed that one of the foundation breeds of the Belgian Hare was the Flemish Giant. It has been said that 6,000 Belgian Hares were imported to the United States in the year 1900 alone, with the breed being known as "The Business Rabbit of the World." Despite the massive interest and huge popularity during the peak period of Belgian Hare popularity (1898–1901), interest in the breed dwindled quickly after that and they are not a particularly common breed of rabbit today. They do, however, retain the historical significance of having boosted American interest in domestic rabbits in a huge way.

Size: Medium

Shape/type: Full arch

Colors: Rich red, tan or chestnut, body ticking, black/blue undercolor

Weight:
Ideal buck: 8 pounds
Ideal doe: 8 pounds
Registration buck: 6 to 9 ½ pounds
Registration doe: 6 to 9 ½ pounds

ALBC status: Threatened (estimated global population of fewer than 1,000)

Beveren

Also known as: The Breed of Distinction

Breed description: Once one of the popular "blue breeds," the Beveren was developed in the small town of Beveren, Belgium, near Antwerp. The Beveren was considered one of the ultimate fur rabbits in the early 1900s, with the lavender shade of blue coloring being the most preferred. As also happened with another predominantly blue breed, the American, the litters of blue Beverens began to contain the occasional white Beveren kit. The first of these sports (throwbacks) is said to have been noted in 1916. Today the Whites are a recognized variety of Beveren and distinguished by their unique blue eyes. In addition, the ARBA recognizes the Black variety of Beveren rabbit and the British Rabbit Council recognizes Brown- and Lilac-colored Beverens. The Brown variety has a certificate of development with ARBA, so it might become an accepted variety in the United States within a few years. Interestingly enough, the forerunner to the British Rabbit Council was the Beveren Club, which was formed in 1918 in Birmingham. Subsequent name changes and mergers with other organizations have resulted in the British Rabbit Club of today, but its original roots were with the Beveren organization.

Size: Large

Shape/type: Semi-arch

Colors: Black, Blue, and White (Blue-Eyed)

Weight:
Ideal buck: 10 pounds
Ideal doe: 11 pounds
Registration buck: 8 to 11 pounds
Registration doe: 9 to 12 pounds

ALBC status: Watch (estimated global population of less than 2,000)

Breed specialty club website: www.freewebs.com/beverens

Britannia Petite

Also known as: The Fancy's Elite

Breed description: The Britannia Petite is actually the North American name for a British rabbit breed, the Polish. Since the ARBA already recognized a different rabbit breed as the Polish, these British imports were given the name of Britannia Petites. This descriptive name aptly recognizes the breed's country of origin (Great Britain), as well as its size (petite). These diminutive rabbits have the appearance of a small hare, due to their full-arch type. The Ruby-Eyed White was the preferred and desired color of the original imports to the United States, and all of the initially imported Britannia Petites were this color. Although the Ruby-Eyed White was the only color recognized by the ARBA in the 1970s and 1980s, other colors have subsequently been added to the recognized list in recent years. Additional colors are in the process of being presented for possible ARBA approval, including Smoke Pearl Marten.

Size: Small

Shape/type: Full arch

Colors: Black, Black Otter, Chestnut, Sable Marten, Ruby-Eyed White

Weight:
Registration buck: Up to 2 ½ pounds, maximum
Registration doe: Up to 2 ½ pounds, maximum

ALBC status: None

Breed specialty club website:
www.britanniapetites.com

Californian

Also known as: From East Coast to West, Californians are the Best

Breed description: Although named because it (obviously) originated in California, the Californian breed of rabbit was originally called the Cochinelles. However, its Californian heritage gave way to the eventual name of this popular breed. In the early 1920s, George West began experimenting with the production of a large commercial rabbit, bred specifically as a meat rabbit and for its fur. He began by crossing Himalayan rabbits with Standard Chinchillas and eventually produced a Chinchilla-colored half-bred buck, which he subsequently bred to New Zealand White does. At this point, he began reaching his desired size and fur goals, along with the Himalayan coloring. With the help of two other dedicated California breeders, West perfected the Californian breed and the breed received its working standard from ARBA in 1939. Today it is one of the most popular commercial rabbits, in addition to having the advantage of its striking coloring.

Size: Large

Shape/type: Commercial

Colors: White with black points (ears, nose, feet, and tail)

Weight:
Ideal buck: 9 pounds
Ideal doe: 9 ½ pounds
Registration buck: 8 to 10 pounds
Registration doe: 8 ½ to 10 ½ pounds

ALBC status: None

Breed specialty club website:
www.nationalcalclub.com

Champagne d'Argent

Also known as: The Silver Beauty

Breed description: The Champagne d'Argent is a French breed whose name means "Silver Rabbit from Champagne" in French. It is known as the Argente de Champagne in France and has been bred for its fur in France for many years, some sources state as early as 1631. Exportations became common in the early 1900s, with many rabbits being exported to Japan, Belgium, Germany, England, and the United States. The first Champagne d'Argent rabbits were brought to the United States in 1912 and were called Champagne Silver or French Silver, but the breed remained rare in this country for many years. The breed was not exported to England until 1919 but has since become one of the most popular fur breeds in that country. The ideal coloring is that of old silver, a mixture of white, blue, and black with a darker "butterfly" pattern on the nose. Any yellow tinge or coloring on the body of the rabbit is a serious defect.

Size: Large

Shape/type: Commercial

Colors: Only acceptable coloring is Silver, a bluish white with black hairs interspersed, and a dark nose and muzzle. The shades do vary within individual rabbits, but silver is the required color.

Weight:
Ideal buck: 10 pounds
Ideal doe: 10 ½ pounds
Registration buck: 9 to 11 pounds
Registration doe: 9 ½ to 12 pounds

ALBC status: none

Breed specialty club contact information: Champagne d'Argent Rabbit Federation, 1704 Heisel Ave., Pekin, IL 61554

Checkered Giant

Also known as: The Rabbit Beautiful

Breed description: Although the background of the Checkered Giant is somewhat cloudy and hard to pinpoint, there is a general consensus that the breed is German. Some German-bred spotted rabbits were crossed with Flemish Giants to produce the Great German Spotted rabbit (known in Germany as the Deutsche Riesenschecke). In 1904, these were crossed with the Flemish Giant, which resulted in what is now the Checkered Giant (though it was formerly known by several different names, including the German Spotted and American Checkered Giant). The first were imported to the United States in 1910, and the imports during the 1920s and 1930s were larger rabbits. The type in the United States today has altered slightly from the European Checkered Giants. Several colors were acceptable during the early years of the breed; however, since 1950, only Black and Blue are considered acceptable colors. Only one Checkered Giant has ever won the coveted Best in Show at an ARBA convention, and that was in 1976.

Weight:
Registration buck: minimum weight 11 pounds
Registration doe: minimum weight 12 pounds

ALBC status: none

Breed specialty club website: www.acgrc.com

Size: Giant

Shape/type: Full arch

Colors: There are only two recognized colors of Checkered Giants, Blue and Black. The markings must be distinct, including the butterfly, the eye circles, the check spots, ears, spine, tail, and side markings.

American Chinchilla

Breed description: As the name implies, the American Chinchilla is a purely American breed, bred up in size and selectively bred from heavy-weighted Standard Chinchillas. While fur and color are important characteristics to the breed, type is considered more important than color. The breed was originally known as the American Heavyweight Chinchilla. The American Chinchilla achieved a large following during the 1920s when more than 17,000 Chinchillas were registered in one year alone. It is now the least common of the three Chinchilla types and is currently on the American Livestock Breeds Conservancy "critical" list.

Size: Large

Shape/type: Commercial

Colors: Chinchilla (hair base is blue, middle is gray, then white, and black tipped)

Weight:
Registration buck: 9 to 11 pounds
Registration doe: 10 to 12 pounds

ALBC status: Critical (estimated global population of fewer than 500 animals)

Breed specialty club website:
www.acrba.net

Giant Chinchilla

Also known as: The Gentle Giant

Breed description: Like the American Chinchilla, the Giant Chinchilla is another North American creation produced from selective breeding of large Standard Chinchillas with Flemish Giants, New Zealand Whites, Champagne d'Argents, and American Blues. Edward Stahl of Missouri is credited with the production of the first successful Giant Chinchillas and reached his ideal in type and size in a litter born in 1921. His goal was a giant-sized rabbit with excellent commercial value and quality pelt and meat. The Giant Chinchilla received its working standard from the ARBA in 1928, and Stahl (known as the father of the Domestic Rabbit Industry in America) has gone down in history as the first (and so far the only) person to sell a million dollars worth of rabbit-breeding stock.

Weight:
Registration buck: 12 to 15 pounds
Registration doe: 13 to 16 pounds

ALBC status: Watch (estimated global population of fewer than 2,000)

Breed specialty club website:
www.giantchinchilla.com

Size: Giant

Shape/type: Semi-arch

Colors: Chinchilla (hair base is blue, middle is gray, then white, and black tipped)

Standard Chinchilla

Also known as: The Original Chin

Breed description: The standard-sized Chinchilla was developed in France by M. J. Dybowski. He crossed a wild gray rabbit (called a Garenne) on a Himalayan and also crossed the Garenne on a blue rabbit. By mating the offspring of these two crosses, he achieved the Chinchilla coloring. Some sources believe that there was probably an infusion of black-and-tan rabbits at some point. The first Chinchilla rabbit arrived in America in 1919, and the standard was approved in 1924. The Chinchilla has been very popular as a foundation for many of the more recently developed rabbit breeds.

Size: Medium

Shape/type: Compact

Colors: Chinchilla (hair base is blue, middle is gray, then white, and black tipped)

Weight:
Ideal buck: 6 pounds
Ideal doe: 6 ½ pounds
Registration buck: 5 to 7 pounds
Registration doe: 5 ½ to 7 ½ pounds

ALBC status: None

Breed specialty club contact information: American Standard Chinchilla Rabbit Breeders Association, 1607 9th Street West, Palmetto, FL 34221

Cinnamon

Also known as: The Spice of the Rabbits

Breed description: Aptly named for its unique cinnamon shade of coloring, this large meat-type rabbit originated in Montana and was developed by a 4-H family. The coloring was the unexpected result of the mating of a New Zealand White/Chinchilla mix buck and a Checkered Giant/Californian mix doe. The resulting litters had the occasional rust- or cinnamon-colored kit, and one litter produced a pair of these Cinnamon kits. The family began breeding specifically for this coloring in a commercial-type rabbit. The breed had its first presentation in 1969 at the ARBA convention in Calgary, Canada, and achieved recognition in 1972. The Cinnamon is not a common breed in the United States, yet there were 29 of them exhibited at the 2006 ARBA convention in Fort Worth, Texas.

Size: Large

Shape/type: Commercial

Colors: Rust- or cinnamon-colored with gray ticking, butterfly nose, eye circles, hind leg spots

Weight:
Ideal buck: 9 ½ pounds
Ideal doe: 10 pounds
Registration buck: 8 ½ to 10 ½ pounds
Registration doe: 9 to 11 pounds

ALBC status: None

Breed specialty club contact information: Cinnamon Rabbit Breeders Association, c/o Nancy Searle, 550 Amherst Rd., Belchertown, MA 01007

Creme d'Argent

Also known ss: The Cream of the Fancy

Breed description: Like the Champagne d'Argent, the Creme d'Argent originated in France, and has a long history in that country. Its popularity stems from its unique fur coloring, an orange-silver color, which is the most important characteristic of the breed. The first Creme d'Argents were imported to the United States in the 1920s and 1930s, although the litters produced from these original imports did not always reproduce the desired orange-silver coloring. To achieve the proper color, an outcross to a Fawn Flemish rabbit was successfully undertaken. Today, breeders are still working to maintain the orange-silver coloring and are carefully breeding and culling undesirable shades or hues. The Creme d'Argents in the United States have been selectively bred to be larger than those bred in Europe, which has caused some difficulties in the body shape and shoulder slope of the Creme d'Argents bred in the United States.

Size: Large

Shape/type: Commercial

Colors: Cream with orange undercoat, butterfly marking

Weight:
Ideal buck: 9 pounds
Ideal doe: 10 pounds
Registration buck: 8 to 10 ½ pounds
Registration doe: 8 ½ to 11 pounds

ALBC status: Watch (estimated global population of fewer than 2,000)

Breed specialty club website:
www.cremedargentfederation.com

Dutch

Also known as: World's Finest Show Rabbit

Breed description: A very old breed whose ancestors hailed from Holland, the Dutch rabbit is an extremely popular fancy rabbit, as it is judged primarily on its markings. The breed dates to the early 1800s, and by the late 1800s, clubs were being formed in Europe. The Dutch is one of the first breeds that the National Pet Stock Association (the forerunner to ARBA) recognized. Dutch rabbits are one of the most easily recognized rabbit breeds, due to their unique style of markings: a solid-colored body with white markings on the neck, feet, and a blaze face. The body color can be any of several colors, coupled with the white markings. Like the English Spot, the Dutch rabbit is truly an enjoyable breed for the rabbit enthusiast who enjoys the challenge of attempting to produce a perfectly marked specimen.

Size: Small

Shape/type: Compact

Colors: White with markings of Black, Blue, Chocolate, Gray, Steel, or Tortoise

Weight:
Ideal buck: 4 ½ pounds
Ideal doe: 4 ½ pounds
Registration buck: 3 ½ to 5 ½ pounds
Registration doe: 3 ½ to 5 ½ pounds

ALBC status: None

Breed specialty club website:
www.dutchrabbit.com

Dwarf Hotot

Also known as: The Eyes of the Fancy

Breed description: This unique and diminutive breed is actually a dwarf version of the standard-sized Hotot rabbit. The Dwarf Hotot was developed in Germany by two different breeders in the 1970s. One breeder crossed black and white Netherland Dwarfs and occasionally a Netherland Dwarf kit was born with Hotot markings. The other breeder crossed a standard-sized Hotot with a Ruby-Eyed White Netherland Dwarf and began producing Hotot dwarfs. Interestingly enough, Dutch markings occasionally appeared during these attempts at creating Dwarf Hotots. The breed was first presented at an ARBA convention in 1981 and was approved in 1984 as its own breed, not a variant color of a Netherland Dwarf.

Size: Small

Shape/type: Compact

Colors: White with black- or chocolate-colored circles around the eyes

Weight:
Ideal buck: 2 ½ pounds
Ideal doe: 2 ½ pounds
Registration buck: Up to 3 pounds
Registration doe: Up to 3 pounds

ALBC status: None

Breed specialty club website:
www.adhrc.com

English Spot

Also known as: Spotted Beauty of the Rabbitdom

Breed description: Established in England during the early 1800s, the English Spot began earning recognition as its own breed in the 1890s, though it was sometimes called the English Butterfly or the Papillon (which means butterfly in French). Interest in the United States first occurred after the Belgian Hare enthusiasm began waning after 1901. The breed is admired by many as the ultimate in fancy rabbits, probably because producing a perfectly marked specimen is a difficult undertaking, which makes it all the more rewarding when an ideal English Spot is produced. In every litter there is the possibility of producing a rabbit with proper markings, a rabbit that is solid (self) colored, and a rabbit with incorrect marking (known as a "Charlie"). The perfect markings include chain markings, butterfly nose, eye circles, and cheek spots.

Size: Medium

Shape/type: Full arch

Colors: White with markings in the following colors: Black, Blue, Chocolate, Gold, Gray, Lilac, and Tortoise

Weight:
Ideal buck: 6 pounds
Ideal doe: 7 pounds
Registration buck: 5 to 8 pounds
Registration doe: 5 to 8 pounds

ALBC status: None

Breed specialty club website:
www.englishspots.8m.com

Flemish Giant

Also known as: The Universal Breed

Breed description: The history of the Flemish Giant is somewhat of a mystery, with some disagreement by experts as to the definitive source of its origin. However, it is universally acknowledged that the Flemish Giant has been an established breed for several hundred years. Some sources believe that the Flemish Giant originated in Flanders, while others feel that it may have originally come from China. The first Flemish Giants were brought to the United States in the 1890s. The original colors recognized by the ARBA were the Black, Light Gray, and Steel Gray varieties, followed by the Blue and White varieties. Later on came the Sandy coloring, followed in 1938 by the Fawn variety. Sandy is now the most popular color in the United States; however, the only color recognized by the British Rabbit Council is Steel Gray.

Size: Giant

Shape/type: Semi-arch

Colors: Black, Blue, Fawn, Light Gray, Sandy, Steel Gray, White

Weight:
Registration buck: 13 pounds and over
Registration doe: 14 pounds and over

ALBC status: None

Breed specialty club website:
www.nffgrb.com

Florida White

Also known as: The Little White Rabbit That Fills So Many Needs

Breed description: The Florida White breed of rabbit is aptly named; they are white and originated in Florida. The original proponent of the breed was Orville Miliken of Florida, a rabbit judge who became interested in producing a small white rabbit as a laboratory animal and small meat rabbit. He used white examples of Polish and Dutch rabbits, as well as a New Zealand White rabbit in his breeding program during the 1960s. In 1967, the Florida White was recognized by the ARBA. Subsequent breeders worked toward improving and perfecting the breed, and the 1999 ARBA National Convention Best in Show rabbit was none other than a Florida White.

Size: Small

Shape/type: Compact

Colors: White with pink eyes

Weight:
Ideal buck: 5 pounds
Ideal doe: 5 pounds
Registration buck: 4 to 6 pounds
Registration doe: 4 to 6 pounds

ALBC status: None

Breed specialty club website:
www.geocities.com/fwrba

Harlequin

Also known as: The Royal Jester

Breed description: Although there is some question as to the exact origin of the colorful Harlequin rabbit, most sources believe it to be of French origin and possibly from a cross of a tri-colored Dutch rabbit and a "backyard" rabbit. Other sources believe that it may have a Japanese origin. That possibility is intriguing when considering that the original name of the breed was the Japanese. The breed's first exhibition was in France in 1887 and the unique calico coloring made it very interesting to rabbit fanciers. The first Japanese rabbits were imported to the United States in 1919, although the name was changed in the United States and England to Harlequin during World War II. However, the term *Japanese* is still used to describe one of the color varieties of the Harlequin breed (the other variety being Magpie).

ALBC status: None

Breed specialty club website:
www.geocities.com/~harlies

Size: Medium

Shape/type: Commercial

Colors: Black, Blue, Chocolate, and Lilac; banded with orange in the Japanese variety; banded with white in the Magpie variety

Weight:
Ideal buck: 7 ½ pounds
Ideal doe: 8 pounds
Registration buck: 6 ½ to 9 pounds
Registration doe: 7 to 9 ½ pounds

Havana

Also known as: The Mink of the Rabbit Family

Breed description: The first of what is now known as the Havana breed appeared in a litter of common Dutch-marked rabbits in Holland in 1898. The rich brown coloring was reminiscent of the color of a Havana cigar, and it is believed that this may have been the origin of the breed's name. The first Havanas came to the United States in 1916 and the Chocolate variety was the only coloring recognized. However, after being presented with multiple litters that included blue sports, a man named Lee Owen Stamm began working on breeding a new Blue variety of Havana during the 1950s and 1960s. He is also responsible for the Black variety, which was recognized in 1980. There have also been certificates of development issued for two additional varieties of Havana (Broken and Lilac), so those may be accepted colors in the near future. The Havana is also recognized as being the breed where the mutated Satin fur variety first appeared. This immensely popular Satin fur is now been selectively bred and developed into three different breeds: Satin, Mini Satin, and Satin Angora.

Size: Medium

Shape/type: Compact

Colors: Black, Blue, Chocolate

Weight:
Ideal buck: 5 ¼ to 5 ½ pounds
Ideal doe: 5 ¼ to 5 ½ pounds
Registration buck: 4 ½ to 6 ½ pounds
Registration doe: 4 ½ to 6 ½ pounds

ALBC status: None

Breed specialty club website:
www.havanarb.com

Himalayan

Breed description: One of the oldest recognized rabbit breeds in the world, the Himalayan has an origin that remains a mystery. Some experts believe that the Himalayan rabbit is of Chinese origin, while other experts say that the breed was developed around the Himalayan Mountain region for which the breed is named. In any case, the ARBA says the Himalayan has a "wider distribution throughout the world than any other rabbit." The National Pet Stock Association (predecessor to ARBA) granted recognition for the Himalayan in 1910, although for only the Black coloring. The additional varieties of Blue, Chocolate, and Lilac were added in subsequent years. The Himalayan breed is the only ARBA-recognized breed to have the cylindrical body type. Enthusiasts of several other rabbit breeds have worked to introduce the striking Himalayan coloring into their breeds.

Size: Small

Shape/type: Cylindrical

Colors: White with dark points (Black, Blue, Chocolate, and Lilac are recognized colors), ruby-red eyes

Weight:
Ideal buck: 3 ½ pounds
Ideal doe: 3 ½ pounds
Registration buck: 2 ½ to 4 ½ pounds
Registration doe: 2 ½ to 4 ½ pounds

ALBC status: None

Breed specialty club website:
www.ahra.himmie.net

Hotot

Breed description: Originally known and recognized by the ARBA as the Blanc de Hotot rabbit, today the breed is known simply as the Hotot. Its distinctive coloring is unique among rabbit breeds, although the breed's miniature version, the Dwarf Hotot also shares the coloring. The rabbits are entirely white with black eyes and bands of black around the eyes. They were originally developed in the early 1900s by Madame Eugenie Bernhard of France, who worked for more than a decade toward producing a large white rabbit with black eyes as a show rabbit and for meat and fur. She was somewhat vague regarding the breeds used in the development of the Blanc de Hotots, referring to her original breeding stock as "French spotted rabbits" with differentiation to breed. However, many believe that the Checkered Giant was the foundation of the Blanc de Hotot. They were first exhibited in Paris in 1920, yet it took another 58 years for them to be imported to the United States, and they were recognized by the ARBA shortly thereafter.

Size: Large

Shape/type: Commercial

Colors: White, with black around the eyes (1/16 to 1/8 inch of black)

Weight:
Ideal buck: 9 pounds
Ideal doe: 10 pounds
Registration buck: 8 to 10 pounds
Registration doe: 9 to 11 pounds

ALBC status: Threatened (estimated global population of fewer than 1,000)

Breed specialty club website:
www.geocities.com/blancdehototclub

Jersey Wooly

Also known as: The Fluff of the Fancy

Breed description: Developed in New Jersey during the 1970s, the Jersey Wooly breed was the result of an idea conceived by rabbit breeder Bonnie Seeley. She wanted to produce a dwarf-sized rabbit with Angora wool, feeling that it would be a marketable type of rabbit. She began crossing Angoras with Netherland Dwarf rabbits and her result was a dwarf-sized rabbit with wool similar to a French Angora. She presented the first Jersey Woolies at the 1984 ARBA national convention. Their official recognition was delayed due to mismatched toenails in some of the presentation animals, but the breed eventually achieved its ARBA recognition at the 1988 convention in Madison, Wisconsin. It is now an extremely popular breed enjoyed by many enthusiasts.

Size: Small

Shape/type: Compact

Colors: Agouti, any other variety, Self, Shaded, Tan pattern

Weight:
Ideal buck: 3 pounds
Ideal doe: 3 pounds
Registration buck: Not more than 3 ½ pounds
Registration doe: Not more than 3 ½ pounds

ALBC status: None

Breed specialty club website:
www.njwrc.net

Lilac

Breed description: Although lilac coloring appears in many types of rabbit breeds, the Lilac breed is a medium-sized rabbit with characteristics reminiscent of the Havana breed. Not surprisingly, the Lilacs originated as sports in Havana litters (brown rabbits diluted to lilac) and were originally noted during the early 1900s in England. A handful of breeders began breeding specifically for the Lilac coloring, either by using sports from Havana litters, or by crossing Havana Browns with blue rabbits, such as the Blue Beveren. Nowadays, we have the pinkish gray Lilacs with a uniform Havana type. They have never been hugely popular in the United States, but they do have a dedicated following.

Size: Medium

Shape/type: Compact

Colors: Lilac

Weight:
Ideal buck: 6 to 7 pounds
Ideal doe: 6 ½ to 7 ½ pounds
Registration buck: 5 ½ to 7 ½ pounds
Registration doe: 6 to 8 pounds

ALBC status: Watch (estimated global population of fewer than 2,000)

Breed specialty club website:
www.geocities.com/nlrca2002

English Lop

Also known as: King of the Fancy

Breed description: Believed to have originated in North Africa, the English Lop was brought to Europe, probably in the first half of the nineteenth century. Initially, the breeders of English Lops placed their main emphasis on ear length. Because of the intense concentration on this single feature, the overall quality of the English Lops decreased. Due to the efforts of dedicated breeders and enthusiasts, the breed has recovered and is now an extremely popular fancy rabbit. Today, the ear length must be at least 21 inches, but rabbits with ear lengths of more than 30 inches have been recorded.

Size: Large

Shape/type: Semi-arch

Colors: Agouti, Broken, Self, Shaded, Ticked, Wide Band

Weight:
Registration buck: 9 pounds or more
Registration doe: 10 pounds or more

ALBC status: None

Breed specialty club website:
www.lrca.net

French Lop

Also known as: King of the Fancy

Breed description: It is understood that the French Lop was originally created by crosses of English Lops and Flemish Giants. In the case of the French Lop, the emphasis has not been on ear length as it has been for the English Lops. Instead, breeders sought to develop a larger lop-eared rabbit, hence the use of Flemish Giants in the original breeding programs. French Lops were one of the first breeds recognized in the United States. While historically they have not been as common as the English Lops in the United States, interest is increasing.

Size: Large

Colors: Agouti, Broken, Self, Shaded, Ticked, Wide Band groups

Shape/type: Commercial

Weight:
Registration buck: 10 ½ pounds or more
Registration doe: 11 pounds or more

ALBC status: None

Breed specialty club website: www.lrca.net

Holland Lop

Also known as: The Hallmark Breed

Breed description: Developed during the 1950s and 1960s in the Netherlands by Adrian DeCock, the Holland Lop was created through experimental crosses of French and English Lops with Netherland Dwarf rabbits. They were first brought to the United States in 1976 and were accepted by the ARBA in 1980. The Holland Lop is a hugely popular breed all over the world and enjoyed for the diminutive size of the Netherland Dwarf, combined with the fancy ear type of the French Lop. They are popular show rabbits and are found in more than 100 color varieties, which makes them an enjoyable breed for color enthusiasts.

Size: Small

Colors: Agouti, Broken, Pointed White, Self, Shaded, Ticked, Wide Band

Shape/type: Compact

Weight:
Registration buck: No more than 4 pounds
Registration doe: No more than 4 pounds

ALBC status: None

Breed specialty club website:
www.hlsrc.com

Mini Lop

Breed description: The Mini Lop breed originated in Germany, where they are known as Klein Widder rabbits. They were developed in the mid-twentieth century from a mixture of several breeds, including the French and English Lops, Chinchilla, and Polish rabbits. The first Klein Widders were brought to the United States in 1972, but they did not initially excite much interest in this country. The proponents of the breed decided to change the name of the breed to "Mini Lop," and in 1980 the ARBA officially recognized the breed.

Size: Medium

Shape/type: Compact

Colors: Agouti, Broken, Pointed White, Self, Shaded

Weight:
Ideal buck: 6 pounds
Ideal doe: 6 pounds
Registration buck: 4 ½ to 6 ½ pounds
Registration doe: 4 ½ to 6 ½ pounds

ALBC status: None

Breed specialty club website:
www.minilop.org

Netherland Dwarf

Also known as: The Gem of the Fancy

Breed description: Tiny but adorable, the Netherland Dwarf breed is believed to be the result of an unexpected cross between a Polish rabbit and a wild rabbit, during the early 1900s in Holland. The original Netherland Dwarfs were Ruby-Eyed White and Blue-Eyed White only, but breeders began working on incorporating additional colors into the breed. This pursuit was interrupted by World War II, which devastated Netherland Dwarf rabbit breeding in Europe. However, as seen from the list of ARBA-recognized colors for Netherland Dwarfs, the breed has made a remarkable recovery and is also available in a vast array of colors! The Netherland Dwarf was originally recognized by ARBA in 1969, and since then has become one of the most popular fancy rabbits, with a strong following and support.

Size: Small

Shape/type: Compact

Colors: Black, Blue, Chocolate, Lilac, Blue-Eyed White, Ruby-Eyed White, Sable Point, Siamese Sable, Siamese Smoke Pearl, Tortoise Shell, Chestnut Agouti, Chinchilla Agouti, Lynx Agouti, Opal Agouti, Squirrel Agouti, Tan, Sable Marten, Silver Marten, Smoke Pearl Marten, Otter, Fawn, Himalayan, Orange, Steel, and Broken

Weight:
Ideal buck: 2 pounds
Ideal doe: 2 pounds
Registration buck: Not more than 2 ½ pounds
Registration doe: Not more than 2 ½ pounds

ALBC status: None

New Zealand

Also known as: The Breed in the Lead

Breed description: Although the name implies differently, the New Zealand rabbit was developed in the United States with the first recognized being the Red variety around 1912. Experts believe that Belgian Hares were probably used in the development of the New Zealand Red, possibly crossed with fawn- or white-colored rabbits. The New Zealand White, which is now the most popular variety, first appeared in 1917 in California and was probably produced by crossing the New Zealand Red with the Flemish Giant, Angora, or American Whites. The White variety was accepted by the ARBA in the 1920s, although the Black variety was not recognized until the 1950s. There is a Blue variety, though it is not an officially recognized color by ARBA; however, breeders are working toward having a Broken variety recognized by ARBA.

Size: Large

Shape/type: Commercial

Colors: Black, Red, and White

Weight:
Ideal buck: 10 pounds
Ideal doe: 11 pounds
Registration buck: 9 to 11 pounds
Registration doe: 10 to 12 pounds

ALBC status: None

Breed specialty club website:
www.newzealandrabbitclub.com

Palomino

Breed description: Another truly American creation, the Palomino rabbit was developed in Washington by Mark Young during the 1940s. As the name suggests, the Palomino is golden-colored, an important characteristic of the breed. Young developed the breed by using rabbits of unknown breeding selected for their meat and coloring. He bred selectively for a light brown, golden tan coloring and eventually began obtaining litters with a high percentage of the desired shade. The Palomino rabbits were presented for the first time at the 1952 ARBA convention, but they were not approved until 1957 (Lynx variety) and 1958 (Golden variety).

Size: Large

Shape/type: Commercial

Colors: Lynx, Golden

Weight:
Ideal buck: 9 pounds
Ideal doe: 10 pounds
Registration buck: 8 to 10 pounds
Registration doe: 9 to 11 pounds

ALBC status: None

Breed specialty club website:
www.geocities.com/Petsburgh/Park/4198

Polish

Also known as: The Little Aristocrat

Breed description: Despite what you might infer from the name, the Polish rabbit is not believed to hail from Poland. Some of the earliest recorded information about Polish rabbits shows that they were first shown in England in 1884 and were developed as fancy show rabbits. The breed was one of the earliest imported to the United States and was recognized by the National Pet Stock Association (ARBA's predecessor) in 1912. Originally, only the Ruby-Eyed White variety was recognized, with the Blue-Eyed White recognized in 1938. Additional varieties have been added over the ensuing years, including Black, Chocolate, and Blue. The newest variety recognized has been the Broken variety. Breeders are currently working toward having Himalayan and Lilac varieties approved by ARBA.

Size: Small

Colors: Black, Blue, Broken, Chocolate, Blue-Eyed White, and Ruby-Eyed White

Shape/type: Compact

Weight:
Ideal buck: 2 ½ pounds
Ideal doe: 2 ½ pounds
Registration buck: Not more than 3 ½ pounds
Registration doe: Not more than 3 ½ pounds

ALBC status: None

Breed specialty club website:
www.polishrabbits.com

Rex

Also known as: The King of the Rabbits

Breed description: The first Rex rabbits appeared unexpectedly in litters at the farm of Desire Caillion in France in 1919. These odd, short-coated rabbits were considered sports, yet when these rabbits were mated, it was found that the short coat did occasionally reproduce itself in subsequent litters. Suspecting that the unique fur variety would make for a very marketable breed of rabbit, the Rex continued to be produced. However, due to the emphasis on reproducing the Rex fur, some of the other characteristics of the rabbits were overlooked. In those early days, it was the low quality of the rabbits that caused some breeders to refer to them as "wrecks" rabbits. Fortunately, thanks to decades of careful breeding and selection, the Rex rabbits of today are of a much higher quality than their wrecks-like ancestors. The rabbits were first exhibited in France in 1924, which was the same year that the first Rex rabbits reached the United States. They have long been an extremely popular breed and are highly admired for their quality fur.

Size: Medium

Shape/type: Commercial

Colors: Black, Black Otter, Blue, Broken, Californian, Castor, Chinchilla, Chocolate, Lilac, Lynx, Opal, Red, Sable, Seal, White

Weight:
Ideal buck: 8 pounds
Ideal doe: 9 pounds
Registration buck: 7 ½ to 9 ½ pounds
Registration doe: 8 to 10 ½ pounds

ALBC status: None

Breed specialty club website:
www.nationalrexrc.com

Mini Rex

Also known as: The Heir to the Throne

Breed description: One of the newer breeds recognized by the ARBA, the Mini Rex has achieved immense popularity in a relatively short period of time. The breed was accepted by the ARBA in 1988 and was initially created through the use of Dwarf Rex and Rex rabbits. The Dwarf Rex is a European variety of a miniature Rex rabbit and contains a good measure of Netherland Dwarf breeding behind them. The Dwarf Rex is not recognized by the ARBA. The popularity of the Mini Rex is credited to its appealing small size and the added desirability of the Rex fur characteristics. They are also enjoyed by breeders who like working with a breed that is recognized in a multitude of colors.

Size: Small

Shape/type: Compact

Colors: Black, Blue, Broken, Castor, Chinchilla, Chocolate, Himalayan, Lilac, Lynx, Opal, Red, Sable Point Seal, Tortoiseshell, and White

Weight:
Ideal buck: 4 pounds
Ideal doe: 4 ¼ pounds
Registration buck: 3 to 4 ¼ pounds
Registration doe: 3 ¼ to 4 ½ pounds

ALBC status: None

Breed specialty club website:
www.nmrrc.com

Rhinelander

Also known as: Calico of the Fancy

Breed description: Of German origin, the Rhinelander breed of rabbit was developed in 1901 and was essentially achieved by crossing English Spot rabbits with Japanese Harlequins. This resulted in a rabbit with the tri-colors of the Harlequin and the coat pattern of the English Spots. They were originally brought to the United States in 1923, but interest lagged and the breed neared extinction in the United States for several decades. Interest rebounded in the 1970s, and the breed received ARBA status in 1975. The Rhinelander is judged mainly on type, color, and markings. While the Black-and-Tan variety is currently the only recognized coloring in the United States, there is also interest in having the Blue-and-Fawn variety approved, as it is already recognized in some parts of Europe. A certificate of development for the Blue-and-Fawn variety has been issued, but it is still several years from reaching recognized status.

Size: Large

Shape/type: Full arch

Colors: White with spots (essentially a tri-colored version of the English Spot). It must include butterfly and spine markings, colored ears, cheek spots, and eye circles.

Weight:
Ideal buck: 8 pounds
Ideal doe: 8 ½ pounds
Registration buck: 6 ½ to 9 ½ pounds
Registration doe: 7 to 10 pounds

ALBC status: Watch (estimated global population of fewer than 2,000)

Breed specialty club website:
hop.to/Rhinelanders

Satin

Also known as: The Rabbit of Beauty and Distinction

Breed description: Imagine the surprise of a Havana rabbit breeder in the 1930s when he discovered a strange thing about one of his litters. Some of the kits possessed a uniquely reflective coat sheen that had never been noted in previous litters. Genetic testing proved that the unusual fur type was the result of a genetic mutation. Because the Satin fur is caused by a recessive gene, breeders have been able to introduce Satin fur into a variety of other breeds and firmly establish the type in the Satin breed itself. They were originally shown in Havana classes against other Havanas with regular fur, but this resulted in immense displeasure from Havana breeders and the Satin rabbits came to be known as Satin Havanas with their own classes, and ultimately, their own breed. Today, 11 colors of Satin rabbits are recognized by ARBA.

Size: Large

Shape/type: Commercial

Colors: Black, Blue, Broken, Californian, Chinchilla, Chocolate, Copper, Otter, Red, Siamese, White

Weight:
Ideal buck: 9 ½ pounds
Ideal doe: 10 pounds
Registration buck: 8 ½ to 10 ½ pounds
Registration doe: 9 to 11 pounds

ALBC status: None

Breed specialty club website:
www.asrba.com

Mini Satin

Breed description: One of the newest additions to the ARBA's recognized list of rabbits, the Mini Satin received its official approval at the October 2005 ARBA convention. The breed has been in development in the United States since the early 1990s, but the first successful presentation was at the 2003 ARBA convention. As the name implies, the Mini Satin is a smaller version of the Satin breed with the same brilliant coat luster. It's a Satin in a smaller package and is actually more reminiscent of the original size of the Satin breed in the 1930s. The miniature version was achieved through selective crossing of Satin rabbits with Florida Whites and Mini Rex rabbits. Although white is the only variety currently recognized by the ARBA, several additional colors are in various stages of presentation in the approval process, including Red, Copper, Opal, Otter, Black, Chocolate, Himalayan, Siamese, Chinchilla, and Californian.

Size: Small

Shape/type: Commercial

Colors: White (additional colors in the process of approval)

Weight:
Ideal buck: 4 pounds
Ideal doe: 4 pounds
Registration buck: 3 ¼ to 4 ¾ pounds
Registration doe: 3 ¼ to 4 ¾ pounds

ALBC status: None

Breed specialty club website:
www.asrba.com

Silver

Also known as: The Sterling Breed

Breed description: The Silver rabbit is a fancy rabbit judged mainly on its coat color and silvering. It is believed that these rabbits were brought to Europe during the seventeenth century. Crosses with Belgian Hares produced the Brown variety of the Silver rabbit, while Creme D'Argent breeding may be behind the Fawn coloring. The breed was recognized early on in ARBA history, and the first Silver rabbits were brought to the United States in the early 1900s.

Size: Medium

Shape/type: Compact

Colors: Black, Brown, Fawn; all with Silver hair throughout the coat

Weight:
Ideal buck: 6 pounds
Ideal doe: 6 pounds
Registration buck: 4 to 7 pounds
Registration doe: 4 to 7 pounds

ALBC status: Threatened (estimated global population of fewer than 1,000)

Breed specialty club website:
http://natlsilverrabbitclub.tripod.com

Chapter 7

Silver Fox

Also known as: One of a Kind Since 1929

Breed description: Introduced into ARBA in the 1920s, the Silver Fox breed was originally known as the American Heavyweight Silver. After 1929 the breed was renamed the American Silver Fox, and today it is known simply as the Silver Fox, named for its similarity in coloring to a silver fox. They are a larger version of the Silver rabbit and the size increase is due possibly from influence from the Checkered Giants. This breed is found only in North America and only the black variety is currently recognized by ARBA. A blue variety was previously recognized but was dropped from ARBA recognition due to a lack of interest, although there has been a certificate of development issued for the blue variety and it appears that it may again be recognized by ARBA in the future.

Size: Large

Shape/type: Commercial

Colors: Black

Weight:
Ideal buck: 9 ½ pounds
Ideal doe: 10 ½ pounds
Registration buck: 9 to 11 pounds
Registration doe: 10 to 12 pounds

ALBC status: Critical (estimated global population of fewer than 500)

Breed specialty club website:
www.nationalsilverfoxrabbitclub.org

Silver Marten

Breed description: In the 1920s, Chinchilla breeders all over the world began noting the occasional appearance of an oddly colored sport in their litters. These sports became the foundation of the Silver Marten breed, as they are essentially a Black and Tan rabbit with a white gene, rather than a yellow/tan gene. The Black Silver Marten variety was accepted by ARBA in 1929 and the blue variety followed shortly thereafter. The sable variety was accepted in 1934, and the most recently recognized variety is the chocolate.

Size: Large

Shape/type: Commercial

Colors: Black, blue, chocolate, sable

Weight:
Ideal buck: 7 ½ pounds
Ideal doe: 8 ½ pounds
Registration buck: 6 to 8 ½ pounds
Registration doe: 7 to 9 ½ pounds

ALBC status: None

Breed specialty club website:
www.silvermarten.com

Tan

Also known as: Aristocrat of the Fancy

Breed description: A uniquely colored rabbit with a distinctive coat pattern, the Tan rabbits were developed in England during the late 1800s, the first variety being the Black and Tan. They were first exhibited in France in 1894, and the first Black and Tan rabbits were exported to the United States in the early 1900s. In addition to the Black and Tan variety, the ARBA also recognizes three additional colors: the Blue and Tan, Chocolate and Tan, and Lilac and Tan. It is believed that the Blue variety was developed by introducing sooty fawn coloring into the Blacks. The background of the Chocolates remains a mystery, but it is understood that the lilac variety was achieved by the crossing of the Blue and Chocolate varieties.

Size: Small

Shape/type: Full arch

Colors: Black, Blue, Chocolate, Lilac

Weight:
Registration buck: 4 to 5 ½ pounds
Registration doe: 4 to 6 pounds

ALBC status: None

Breed specialty club website:
www.atrsc.org

Thrianta

Breed description: One of the two newest breeds recognized by the ARBA (the other is the Mini Satin), the Thrianta recently achieved its recognition in October 2005 and has only been eligible for full show status since February 2006. The breed originated in the Netherlands in the first half of the twentieth century and was bred specifically for the deep orange coloring, using a mixture of Tan and Havana rabbits. The first Thrianta rabbits came to the United States in the mid-1990s and are attracting a good deal of attention among rabbit fanciers. The coloring is considered the most important characteristic and has the heaviest emphasis (30 points) in judging.

Size: Medium

Shape/type: Compact

Colors: Deep orange red

Weight:
Registration buck: 4 to 6 pounds
Registration doe: 4 to 6 pounds

ALBC status: None

Breed specialty club website:
http://www.geocities.com/thriantarba

Additional Breeds

Although they are not officially recognized by ARBA at the time of this writing, there are additional breeds currently working toward ARBA sanctioning and, therefore, merit mentioning in this text.

Lionhead

Breed description: The Lionhead rabbit is rapidly gaining a great deal of attention among rabbit enthusiasts. These charming little rabbits with the wooly manes (or beards) are believed to have originally been the result of a genetic mutation during the 1960s. They are small in stature and are likely to be a popular breed in the future. They were approved at their first ARBA presentation in 2005, but unfortunately failed their second presentation in 2006. While this is undoubtedly a setback to the breed, the support and enthusiasm by breeders will certainly ensure that the Lionhead will achieve its sanctioning at a later date. Despite the current lack of ARBA sanctioning, many people are already breeding Lionheads in their rabbitries and it is anticipated that interest will only increase as time passes. The North American Lionhead Rabbit Club held a National Exhibition show in 2006 with 605 Lionhead rabbits exhibited. Nearly 100 Lionheads were exhibited at the 2006 ARBA Convention, even though they were unable to show in the sanctioned divisions, they were still shown in exhibition classes.

Size: Small

Shape/type: Compact

Colors: The current working standard allows for only one color of Lionhead rabbit (Tortoiseshell), but additional certificates of development have been issued to cover the following colors: Black, Blue, Chestnut Agouti, Fawn, Ruby-Eyed White, Sable Point, and Siamese Sable.

Weight:
(As described in the working standard)
Ideal buck: 3 ½ pounds
Ideal doe: 3 ½ pounds
Registration buck: 3 ¾ pounds
Registration doe: 3 ¾ pounds

ALBC status: None

Breed specialty club website:
www.lionhead.us

Velveteen Lop

Breed description: The Velveteen Lop is a more recent creation and has only been in development since the 1980s by several breeders in the United States. The Velveteen Lop originally began with a cross between Rex and English Lops in an attempt to obtain a rabbit with the desirable coat qualities of the Rex, with the impressively lopped ears of the English Lop. The process of presenting the Velveteen Lops for ARBA sanctioning has been a long and trying one with many setbacks and a few failed presentations. The breed was scheduled for presentation at the 2006 convention, but was not shown, thus creating another delay in the process of moving forward. Like the Lionhead, there is a great deal of interest in the Velveteen Lop and its future is undoubtedly very bright.

Size: Medium

Shape/type: Semi-arch

Colors: Agouti, Pointed White, Self, Shaded, Ticked, Wide Band

Weight:
(As described in the working standard)
Ideal buck: 5 ¾ pounds
Ideal doe: 5 ¾ pounds
Registration buck: 5 to 6 ½ pounds
Registration doe: 5 to 6 ½ pounds

ALBC status: None

Breed specialty club website:
http://velveteenlop.net

Canadian Plush Lop

Breed description: One of the most recently established breeds, the Canadian Plush Lop has been in development in Canada since 2000 by a breeders group consisting of six rabbitries. The foundation breeds that were used to produce the Canadian Plush Lop are Mini Rex and Holland Lops, with breeders attempting to create a rabbit with the combined features of a rex coat and lop ears, as well as an endearing disposition. In the ensuing years, it was discovered that the offspring of these initial crosses often possessed curly coats. Today, the Canadian Plush Lop has emerged with the desired characteristics of the curly rex coat, lop ears, and the charming temperaments. The rabbits also possess bounding motion, which is reminiscent of the Tan breed, and upright body posture. The breed's working standard has been presented to ARBA and it is believed that a certificate of development will be issued soon, so it is possible that the Canadian Plush Lop will receive its official ARBA recognition within the next few years.

Size: Small

Shape/type: Semi-arch

Colors: Black, Blue, Broken, Sable, Sable Point, Seal, Pointed White, and Tortoiseshell

Weight:
(As described in the working standard)
Registration buck: 3 ½ to 5 pounds
Registration doe: 3 ½ to 4 ½ pounds

ALBC status: None

Index